Cytochrome P-450

Cytochrome P-450

Edited by
Ryo SATO
Osaka University, Osaka
and
Tsuneo OMURA
Kyushu University, Fukuoka

KODANSHA LTD.　ACADEMIC PRESS
Tokyo　　　　　　　New York • San Francisco • London
　　　　　　　　　　A Subsidiary of Harcourt Brace Jovanovich,
　　　　　　　　　　Publishers

◯ KODANSHA SCIENTIFIC BOOKS

Copyright © 1978 by Kodansha Ltd.

All rights reserved

No part of this book may be reproduced in any form, by photostat, microfilm, retrieval system, or any other means, without the written permission of Kodansha Ltd. (except in the case of a brief quotation for review)

I.S.B.N 0-12-619850-0
Library of Congress Catalog Card Number 78-62317

Co-published by
KODANSHA LTD.
12-21, Otowa 2-chome, Bunkyo-ku, Tokyo 112, Japan
and
ACADEMIC PRESS, INC.
111 Fifth Avenue, New York, N.Y. 10003
ACADEMIC PRESS, INC. (LONDON) LTD.
24/28 Oval Road, London NW1

PRINTED IN JAPAN

Contributors

Numbers in parentheses, indicate the chapter(s) to which the author's contributions submit.

Shigeo HORIE, School of Medicine, Kitasato University, *Sagamihara, Kanagawa 228, Japan* (3.2)

Tetsutaro IIZUKA, Department of Biochemistry, Keio University School of Medicine, *Shinjuku-ku, Tokyo 160, Japan* (3.3)

Yoshio IMAI, Institute for Protein Research, Osaka University, *Suita, Osaka 565, Japan* (3.1)

Yuzuru ISHIMURA, Department of Biochemistry, Keio University School of Medicine, *Shinjuku-ku, Tokyo 160, Japan* (5)

Masayuki KATAGIRI, Department of Chemistry, Faculty of Science, Kanazawa University, *Kanazawa 920, Japan* (4.2, 4.3)

Tsuneo OMURA (editor), Department of Biochemistry, Faculty of Science, Kyushu University, *Hakozaki, Fukuoka 812, Japan* (1, 4.1)

Ryo SATO (editor), Institute for Protein Research, Osaka University, *Suita, Osaka 565, Japan* (2)

Katsuko SUHARA, Department of Chemistry, Faculty of Science, Kanazawa University, *Kanazawa 920, Japan* (4.2)

Kunio TAGAWA, 2nd Department of Biochemistry, Osaka University Medical School, *Kita-ku, Osaka 530, Japan* (4.5)

Shigeki TAKEMORI, Department of Environmental Sciences, Faculty of Integrated Arts and Sciences, Hiroshima University, *Higashisenda cho, Hiroshima 730, Japan* (4.2)

Toshio YAMANO, lst Department of Biochemistry, Osaka University Medical School, *Kita-ku, Osaka 530, Japan* (3.3)

Yuzo YOSHIDA, Faculty of Pharmacy, Mukogawa Women's University, *Nishinomiya 663, Japan* (4.4)

Contents

Preface xi

Chapter 1 Introduction: Short History of Cytochrome P-450
1.1. Discovery of Cytochrome P-450 in Liver Microsomes 1
1.2. Discovery of the Physiological Function of Cytochrome P-450 5
1.3. Wide Distribution of Cytochrome P-450 among Various Forms of Life 7
1.4. Constitution of Cytochrome P-450-containing Oxygenase Systems 8
1.5. Elucidation of the Molecular Properties of Cytochrome P-450 10
1.6. Multiple Forms of Cytochrome P-450 in Liver Microsomes 14
1.7. Present Problems and Future Prospects 16
References 18

Chapter 2 Distribution and Physiological Functions
2.1. Introduction 23
2.2. Distribution of Cytochrome P-450 23
2.3. Catalytic Activities of Cytochrome P-450 26
2.4. Metabolic Roles of Cytochrome P-450 29
References 33

Chapter 3 Molecular Properties
3.1. Purification and Chemical Properties 37
   3.1.1. Purification of Hepatic Microsomal P-450 37
   3.1.2. Chemical Properties 47
   3.1.3. Conversion of P-450 to P-420 54
   3.1.4. Interaction of P-450 with Isocyanides 62
   References 70

3.2. Optical Properties 73
  3.2.1. Development of Research 73
  3.2.2. Difference Spectra 83
  3.2.3. Absolute Absorption Spectra 94
  3.2.4. Extinction Coefficients 100
  3.2.5. Photochemical Action Spectrum of P-450 102
  References 104
3.3. Magnetic Properties 106
  3.3.1. History and an Outline of Microsomal Fe_x 106
  3.3.2. Short Survey of Ligand Field Theory 110
  3.3.3. High Spin P-450 Induced by 3-Methylcholanthrene 115
  3.3.4. EPR Signal Changes of P-450 with Various Ligands or Substrates 118
  3.3.5. EPR Study of P-450 Conversion to P-420 125
  3.3.6. NMR Study of P-450 127
  3.3.7. Magnetic and Molecular Spectroscopic Investigations on P-450 other than EPR and NMR Studies 131
  References 133

Chapter 4 Cytochrome P-450-containing Oxygenase Systems

4.1. Hepatic Microsomal Systems 138
  4.1.1. Electron Transport System of Liver Microsomes 138
  4.1.2. Electron Transfer from NAD(P)H to Cytochrome P-450 in Liver Microsomes 145
  4.1.3. Reconstitution of Oxygenase Activity from Solubilized Microsomal Components 151
  4.1.4. Substrate Specificity of the Hepatic Microsomal Cytochrome P-450 System 155
  References 160
4.2. Adrenocortical Mitochondrial Systems 164
  4.2.1. Sequential Fragmentation of Steroid Hydroxylating Components from Adrenal Cortex 165
  4.2.2. Adrenodoxin 165
  4.2.3. NADPH-Adrenodoxin Reductase 170
  4.2.4. Cytochrome P-450 173
  4.2.5. General Discussion 178
  References 182
4.3. Bacterial Systems 184
  4.3.1. Catalytic Components of the Camphor-Methylene Hydroxylating System (Camphor 5-*exo* Hydroxylase) 186
  4.3.2. Function of $P-450_{CAM}$ 188
  4.3.3. Cytochrome P-450-dependent *n*-Alkane Methyl Hydroxylation (ω-Hydroxylation) 189
  4.3.4. Other Bacterial P-450-type Cytochromes 190

4.3.5. Types of Bacterial Cytochrome P-450-containing Hydroxylase Systems 192
References 193
4.4. Yeast Microsomal Systems 194
4.4.1. General Considerations on Yeast Microsomal Electron Transport Systems 194
4.4.2. Purification and Properties of Yeast Cytochrome P-450 195
4.4.3. Cytochrome P-450-containing Electron Transport Systems of Yeasts 200
4.4.4. Function of Cytochrome P-450-containing Systems of Yeasts 201
References 202
4.5. Induction and Disappearance of Cytochrome P-450 in Yeast Cells 202
4.5.1. Induction of Cytochrome P-450 in Alkane-utilizing Yeast 203
4.5.2. Changes in the Cytochrome P-450 Content of Facultatively Aerobic Yeast Cells Grown under Various Conditions 204
4.5.3. Physiological Significance of Changes in Cytochrome P-450 Content 207
References 208

Chapter 5 Mechanisms of Cytochrome P-450-catalyzed Reactions

5.1. Introduction 209
5.2. Substrate Interaction with Cytochrome P-450 211
5.3. Oxygen Interaction with Cytochrome P-450 214
5.4. The Role of Iron-Sulfer Proteins 216
5.5. Reaction Mechanism of P-450$_{CAM}$ and Reactive Species of Oxygen 220
5.6. Problems in Cytochrome P-450 Systems of Mammalian Tissues 224
References 225

Index 229

Preface

The occurrence in liver microsomes of the CO-binding pigment which is now called cytochrome P-450 was first reported independently by Klingenberg and by Garfinkel in 1958. However, at that time, few people could anticipate the importance of this discovery. In 1959, Tsuneo Omura and I began research on the chemical nature of this curious pigment and after various trials were finally able to demonstrate that it is actually a hemoprotein with unusual properties. Even at that time our feeling was that we had just managed to resolve one small problem in biochemistry. It has therefore been a great surprise and pleasure for us to see the original efforts develop and grow so rapidly over the past 15 years. Today, it is well known that hemoproteins similar in spectral properties to hepatic microsomal cytochrome P-450 are widely distributed in the animal, plant and microbial kingdoms and that they participate as monooxygenases in various biochemical processes of metabolic importance. In animals, cytochrome P-450's are also responsible for oxidative transformations of xenobiotics such as drugs, pesticides, carcinogens, and environmental pollutants. For this reason, cytochrome P-450's are attracting broad attentions among pharmacologists and oncologists as well as biochemists. In addition, many physicochemists have become interested in the anomalous physical properties of this class of hemoproteins.

Despite such increasing interest in cytochrome P-450, no specific monograph on this subject has so far been published, although several good symposium proceedings are available. The purpose of the present book is to document our recent knowledge on cytochrome P-450's, and to render it accessible to those who, for various reasons, wish to gain a broad understanding of this field. It should be noted, however, that not all aspects of cytochrome P-450's are covered in the book. Instead, emphasis is laid on biochemical and physicochemical studies. It is hoped that the information given will provide a sound basis for further enquiries into new aspects of the problem, and that this book will be of lasting value to many people, especially young scientists, who are interested in cytochrome P-450 and related biological problems. Due to the rapidity of progress in this field, some of the contents may already be slightly outdated on publication; however, it is believed that this will not impair the value of the book as a general introductory text.

Finally, I wish to thank the contributing authors for their kind collaboration, and to express my gratitude to Mr. M. Takahatake and Mr. A. J. Smith of Kodansha for their kind assistance in the preparation and promotion of this book.

Osaka

May, 1978

Ryo Sato

CHAPTER 1

Introduction : Short History of Cytochrome P-450[*1]

1.1. Discovery of Cytochrome P-450 in Liver Microsomes
1.2. Discovery of the Physiological Function of Cytochrome P-450
1.3. Wide Distribution of Cytochrome P-450 among Various Forms of Life
1.4. Constitution of Cytochrome P-450-containing Oxygenase Systems
1.5. Elucidation of the Molecular Properties of Cytochrome P-450
1.6. Multiple Forms of Cytochrome P-450 in Liver Microsomes
1.7. Present Problems and Future Prospects

1.1 Discovery of Cytochrome P-450 in Liver Microsomes

Cytochrome P-450 made a quiet debut about 20 years ago. In 1955, G. R. Williams[*2] of the Johnson Foundation for Medical Physics, University of Pennsylvania, first noticed a pigment with a peculiar carbon monoxide-binding spectrum in rat liver microsomes. He was studying the oxidation-reduction kinetics of microsome-bound cytochrome b_5 at the laboratory of B. Chance, using the double-beam recording spectrophotometer designed by Chance and his co-workers which was capable of recording the difference spectra of highly turbid samples, and he observed the appearance of a broad but intense absorption band at 450 nm upon bubbling carbon monoxide into a dithionite-reduced microsomal suspension. His novel observation was not published, but was examined in more detail in the same year by M. Klingenberg[1)] who was also at the same laboratory as a visiting scientist from Germany.

Klingenberg attempted to elucidate the nature of this newly discovered microsomal pigment which could then be detected only by its characteristic carbon monoxide-difference spectrum (Fig. 1.1). The binding of carbon monoxide strongly suggested the presence of a heavy

---

[*1] Tsuneo OMURA, Department of Biochemistry, Faculty of Science, Kyushu University, Hakozaki, Fukuoka 812, Japan
[*2] Cited in a paper by Klingenberg.[1)]

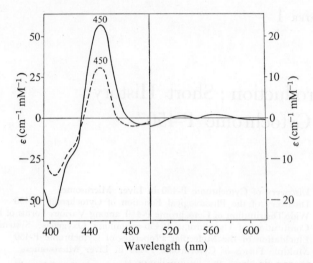

Fig. 1.1. Carbon monoxide-difference spectra for NADH-reduced (- - -) and $Na_2S_2O_4$-reduced (——) rat liver microsomes. The first published record of the spectrum of cytochrome P-450.
(Source: ref. 1. Reproduced by kind permission of Academic Press, Inc., U.S.A.)

metal ion in the chromophore of the pigment, but its spectrum showed no resemblance to any known colored metalloprotein including various hemoproteins. Klingenberg noted that the amount of protohemin in the liver microsomes was more than twice that which could be accounted for by cytochrome b_5, which was the only known hemoprotein in microsomes at the time. However, he was unable to correlate the presence of the excess protohemin in the microsomes with the carbon monoxide-binding pigment because he failed to detect the photo-dissociation of its carbon monoxide compound.

Klingenberg left the Johnson Foundation in the next year, but the study on microsomal redox pigments was continued there by D. Garfinkel[2] who was also greatly interested in elucidating the nature of the "microsomal carbon monoxide-binding pigment". Garfinkel used pig liver microsomes. He attempted to solubilize the pigment, only to find it very unstable to various treatments. The characteristic optical absorption of its carbon monoxide compound at 450 nm disappeared when the microsomes were treated with sodium cholate or pancreatin, and he was unable to obtain a solubilized preparation of this pigment which could be used for elucidating the nature of the pigment. Although the chemical nature of the pigment remained unknown, both the studies of Klingenberg and of Garfinkel were separately published in 1958.[1,2]

No new findings on this peculiar carbon monoxide-binding pigment in liver microsomes were reported in the following years until 1962, when T. Omura and R. Sato of the Institute for Protein Research, Osaka University, published a preliminary account[3] of their studies on this pigment in which they presented conclusive spectral evidence for its hemoprotein nature. They concluded that the pigment was a new b-type cytochrome, and proposed the tentative name "P-450" for it which meant "a pigment with an absorption at 450 nm". Sato was once associated with the Johnson Foundation. He was visiting the Institute when Klingenberg was studying the "microsomal carbon monoxide-binding pigment" at the laboratory of Chance. Thus, the studies of Omura and Sato were also connected with the pioneering work of Williams, Klingenberg, and Garfinkel.

Omura and Sato began work on this subject in 1960. Their first observation which revealed the hemoprotein nature of the pigment was the ethyl isocyanide-difference spectrum of the reduced pigment[3] in which α- and β-bands were clearly noted in addition to intense bands in the Soret region. They also found that the pigment was not wholly destroyed by treatment of microsomes with a detergent, as suggested by previous workers,[1,2] but was quantitatively converted to another spectrally distinct solubilized form which still retained the capacity to combine with carbon monoxide in its reduced state. The carbon monoxide-difference spectrum of the solubilized pigment showed a prominent peak at 420 nm instead of at 450 nm, and it also showed clear α- and β-bands which unquestionably indicated a hemoprotein nature for the solubilized pigment (Fig. 1.2). Furthermore, it was found

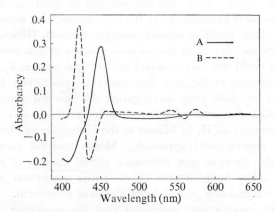

Fig. 1.2. Carbon monoxide-difference spectra of cytochrome P-450 (A) and P-420 (B).
(Source: ref. 4. Reproduced by kind permission of the American Society of Biological Chemists, Inc.)

possible to measure the reduced minus oxidized difference spectrum[3] of the solubilized pigment which was nothing less than a spectrum of a typical b-type cytochrome. The solubilized form was named "P-420" to distinguish it from the native membrane-bound form. Since a quantitative conversion of cytochrome P-450 to P-420 form could be observed when microsomes were treated with a detergent or with a phospholipase preparation, the established hemoprotein nature of the solubilized form, P-420, confirmed that the native membrane-bound form, cytochrome P-450, should also be regarded as a hemoprotein.

A full account of the studies by Omura and Sato on cytochrome P-450 was published in two papers in 1964,[4,5] in which the basic properties of this novel hemoprotein were described. The solubilized form, P-420, was partially purified to remove coexisting colored impurities including cytochrome b_5, and the absolute absorption spectra of P-420 were recorded.[5] Assuming a quantitative conversion of cytochrome P-450 to P-420, they calculated and presented[5] an extinction coefficient value for the carbon monoxide-difference spectrum of reduced cytochrome P-450, which was 91 $cm^{-1} mM^{-1}$ for the absorbance difference between 450 nm and 490 nm. Although the extinction coefficient for cytochrome P-450 was thus first determined by an indirect procedure using microsomes, the validity of this value was later confirmed by determinations with purified cytochrome P-450 preparations.[6] The determination of the extinction coefficient of cytochrome P-450 enabled Omura and Sato to calculate its content in microsomes, and their calculation showed that the combined amount of cytochromes P-450 and b_5 in rabbit liver microsomes was in good agreement with the total amount of protohemin in the microsomes,[4] as separately determined by an alkali pyridine-hemochromogen method. Hence, the presence of the "excess" hemin in microsomes which was first noted by Klingenberg[1] and could not be accounted for by cytochrome b_5, was explained by the presence of the new hemoprotein, cytochrome P-450.

While Omura and Sato were investigating the spectral properties of cytochrome P-450, the same microsomal component was also being studied in the laboratory of H. S. Mason at the University of Oregon[7] using a different experimental approach. Mason and his associates were examining the electron spin resonance (ESR) spectra of rabbit liver microsomes. They found a new characteristic absorption in the $g=2$ region, with a signal at $g_m=2.25$ being the most prominent. They examined the ESR spectra and concluded that the substance giving rise to these signals was probably a low spin ferric hemoprotein, which they designated "microsomal Fe_x". The ESR signal of microsomal Fe_x was significantly reduced when NADH or NADPH was added to microsomes under anaerobic conditions, and the presence of carbon monoxide greatly stimulated the decrease of the signal by NAD(P)H. Moreover, the amount of microsomal Fe_x calculated from the intensity of the signal was almost equal to the difference between the content of total protohemin and that of cytochrome b_5 in the microsomes used. These observations strongly suggested[7] the identity of microsomal Fe_x with cytochrome P-450, which could so far be identified only by its characteristic optical absorption spectra. This suggestion was actually confirmed by later work.[8,9]

1.2 Discovery of the Physiological Function of Cytochrome P-450

The anomalous optical spectra and strong ESR signals of cytochrome P-450, which were both reported in 1962,[3,7] attracted the attention of many biochemists who were engaged in work on the physicochemical properties of hemoproteins. However, an even greater impetus was to be given to the study of this peculiar hemoprotein in the next year. This was the discovery of its physiological function as the oxygen-activating terminal oxidase in various important oxygenation reactions, which was reported by R. W. Estabrook, D. Y. Cooper and O. Rosenthal[10] of the University of Pennsylvania. Utilizing the photochemical action spectrum technique, they demonstrated for the first time the participation of cytochrome P-450 in the steroid C21-hydroxylation reaction catalyzed by adrenal cortex microsomes (Fig. 1.3). They

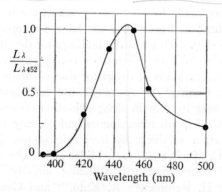

Fig. 1.3. Photochemical action spectrum for the light reversal of carbon monoxide inhibition of steroid C_{21}-hydroxylase activity of adrenal cortex microsomes. The first evidence for the function of cytochrome P-450 as an oxygenase. (Source: ref. 10. Reproduced by kind permission of Springer-Verlag, W. Germany.)

were further able to demonstrate[11] the principal role of cytochrome P-450 in many other mixed function oxidase reactions including hydroxylations or oxidative dealkylations of various drugs by liver microsomes. A review article by Cooper[12] describes how this important discovery was made.

The discovery of the physiological function of cytochrome P-450 opened a vast and expanding field for research on this newly discovered hemoprotein. The photochemical action spectrum for the light reversal of carbon monoxide inhibition of a given oxygenation reaction has become the most reliable standard means for confirming the participation of cytochrome P-450 in the reaction. Not only biochemists, who were naturally much interested in the peculiar optical properties, but also many pharmacologists and endocrinologists began work on cytochrome P-450 since its discovered functions included some of the most important metabolic reactions in their research fields. The increased availability to biochemists of sensitive split-beam recording spectrophotometers also greatly expedited the expansion of studies on cytochrome P-450, since measurement of the absorption spectra of highly turbid samples was essential for studying this hemoprotein. In the mid-1950's, when the first spectral observation of this hemoprotein was made by Williams and others at the Johnson Foundation, few other laboratories in the world were equipped with such a good optical instrument capable of recording accurate absorption spectra of highly turbid suspensions of microsomes or mitochondria.

As further confirmation of the role of cytochrome P-450 in the oxidation of drugs by liver microsomes, brief mention should be made of another important discovery made by pharmacologists at the beginning of the 1960's which contributed greatly to the later development of the work on cytochrome P-450. This was the finding that the drug-oxidizing activities of mammalian liver microsomes could be greatly elevated by administering certain kinds of drugs to the animals (*cf.* the review article by A. H. Conney[13]). Although the hemoprotein nature of cytochrome P-450 and its role in drug oxidation reactions were not elucidated until 1962–63, pharmacologists had already been actively studying the oxidation of drugs by liver microsomes since the pioneering work of G. C. Mueller and J. A. Miller in 1949,[14] and the drug-induced increase of microsomal drug-oxidizing activity was independently discovered by H. Remmer,[15] R. Kato,[16] and Conney *et al.*[17] in 1959–60. As the nature and function of cytochrome P-450 was soon elucidated, the increased drug-oxidizing activity could be explained on the basis of a selective increase in cytochrome P-450 and its reductase, NADPH-cytochrome *c* reductase, in microsomes from drug-treated animals.[18,19] The drug-induced increase of microsomal cytochrome P-450 soon proved very useful for studying this enzyme system as will be described below.

1.3 WIDE DISTRIBUTION OF CYTOCHROME P-450 AMONG VARIOUS FORMS OF LIFE

Since cytochrome P-450 was first found in liver microsomes, it was initially thought to be present only in microsomes of animal tissues.[20] However, the same pigment was soon discovered in the mitochondria of the adrenal gland,[21] and the mitochondrial cytochrome P-450 was identified as the oxygen-activating component of the steroid 11β-hydroxylase[22,23] and side-chain cleavage[24,25] systems which participate in the biosynthesis of adrenocortical steroid hormones from cholesterol. Other animal tissues which produce steroid hormones (corpus luteum, placenta, etc.) were also examined, and the mitochondria obtained from such tissues were invariably found to contain significant amounts of cytochrome P-450.[26,27] The intimate association of cytochrome P-450 with the metabolism of steroids in animal tissues has been further supported by more recent reports suggesting the presence of a cytochrome P-450 in liver mitochondria,[28] where the hemoprotein catalyzes C26-hydroxylation of 5β-cholestane-3α,7α,12α-triol, and also another type of cytochrome P-450 in kidney mitochondria which catalyzes 1-hydroxylation of 25-hydroxy vitamin D_3,[29] although direct spectral demonstration of cytochrome P-450 in these mitochondria has been difficult until recently[30] due to the low concentration of this particular hemoprotein in the presence of high concentrations of cytochromes of the mitochondrial respiratory chain.

In the latter half of the 1960's, successive reports on the discovery of cytochrome P-450 in various forms of life demonstrated that the distribution of this peculiar hemoprotein among living creatures was much wider than initially thought. As early as 1964, A. Lindenmayer and L. Smith[31] described the presence of cytochrome P-450 in the yeast, *Saccharomyces cerevisiae*, and then the first convincing evidence for the presence of this hemoprotein in a bacterial species, nitrogen-fixing *Rhizobium japonicum* bacteroids, was presented by C. A. Appleby[32] in 1967. In contrast with other previously known cytochrome P-450's, which were all bound to membranous subcellular particles, the cytochrome P-450 from *Rhizobium japonicum* bacteroids could be recovered in the supernatant fraction when the bacteroids were mechanically disrupted, and could be partially purified by conventional procedures.[32]

Discoveries of soluble cytochrome P-450's in other bacterial species were then reported successively: a camphor-hydroxylating cytochrome P-450 of *Pseudomonas putida*,[33] a soluble cytochrome P-450 in cell-free extracts of *Corynebacterium* sp.[34] grown with *n*-octane as the sole carbon source, etc. A soluble chloroperoxidase of *Caldariomyces fumago*[35] was also identified from its spectral properties as a kind of cytochrome P-450. Since these bacterial cytochrome P-450's were all soluble, their purification was naturally much easier than that of their membrane-bound counterparts in higher organisms. The cytochrome P-450 of *Pseudomonas putida* was soon highly purified[36] and crystallized,[37] and so became the best material for studying the chemical and physicochemical properties of this peculiar hemoprotein (*cf.* section 4.3).

Cytochrome P-450 has also been identified in the tissues of various vertebrates, invertebrates and plants (*cf.* Chapter 2). In most cases, the cytochrome was found in microsomal fractions obtained from tissue homogenates by fractional centrifugation, and whenever its physiological role was elucidated, it was invariably functional as a monooxygenase utilizing either NADH of NADPH as the source of reducing equivalents for the oxygenation reaction. In the case of plants, the discoveries of cytochrome P-450 were generally associated with the elucidation of the oxidative metabolism of important physiological compounds. That is to say, the 19-hydroxylation of kaur-16-ene by tissue homogenates of *Echinocystis macrocarpa* endosperm,[38] 4-hydroxylation of cinnamic acid by microsomes of *Sorghum bicolor* seedlings,[39] 10-hydroxylation of geraniol or nerol by microsomes of *Vinca rosea* seedlings,[40] etc. were all shown to involve cytochrome P-450. On the other hand, in studies on insects[41-45] and other lower animals,[46-49] greatest attention has so far been paid to the role of cytochrome P-450 in the metabolism of insecticides or other xenobiotic compounds. Undoubtedly, cytochrome P-450 is very widely distributed among various forms of life, from primitive bacteria to highly developed mammals, and participates in diversified metabolic reactions as the oxygen-activating component of monooxygenase systems.

1.4 Constitution of Cytochrome P-450-containing Oxygenase Systems

When the essential role of cytochrome P-450 in various oxygenation reactions had been made clear,[10,11] one of the next important objects of study was the mechanism of supply of electrons from NADPH to cytochrome P-450 where the activation of the oxygen molecule takes place at the expense of the reducing equivalents. The cytochrome P-450-containing oxygenase system of liver microsomes was, however, very tightly associated with the membrane, and resisted early attempts at solubilization. The first cytochrome P-450-containing oxygenase system whose constitution was elucidated was the 11β-hydroxylase system of adrenal cortex mitochondria, in 1965.

In contrast to the microsomal oxygenase system where all components of the system were tightly bound to the membrane, the 11β-hydroxylase system of adrenal cortex mitochondria could be separated by sonic treatment into cytochrome P-450-containing membrane fragments and a soluble NADPH-cytochrome P-450 reductase system,[22,23] and the reductase system was further fractionated by chromatography into an iron-sulfur protein and a flavoprotein which was capable of reducing the iron-sulfur protein by NADPH.[22,23,50-52] All of these three components were required for reconstitution of 11β-hydroxylase activity,[22,23,52] so confirming the essential role of the iron-sulfur protein in the transfer of reducing equivalents from the NADPH-linked flavoprotein to cytochrome P-450. The iron-sulfur protein showed absorption spectra[23,50] similar to those of spinach ferredoxin, and the new name "adrenodoxin" was proposed for it by K. Suzuki and T. Kimura.[51]

A few years later, the involvement of a cytochrome P-450 in the camphor hydroxylase system of *Pseudomonas putida* was discovered.[33] This soluble oxygenase system consisted of a cytochrome P-450, an iron-sulfur protein which was named "putidaredoxin",[53] and a flavoprotein which reduced putidaredoxin by NADH. Thus, an iron-sulfur protein was again found to be functional in a cytochrome P-450-containing oxygenase system as an obligatory intermediate electron carrier between a flavoprotein and cytochrome P-450. This bacterial cytochrome P-450 system was very similar in its constitution to the steroid-hydroxylating cytochrome P-450 system of adrenal cortex mitochondria, although adrenodoxin could not replace putidaredoxin in the reconstituted camphor-hydroxylating system,[54] or *vice versa*.[55]

On the other hand, studies on the cytochrome P-450-containing oxygenase system of liver microsomes revealed that no iron-sulfur protein was involved in it. The system consisted of two protein components, cytochrome P-450 and NADPH-cytochrome *c* reductase which transfers electrons from NADPH to cytochrome P-450. Although the requirement for NADPH in microsomal oxygenation reactions suggested the involvement of an NADPH-specific reductase and various lines of indirect evidence[18,20,56] which had been accumulated by the mid-1960's suggested the participation of NADPH-cytochrome *c* reductase in the supply of reducing equivalents from NADPH to the terminal oxidase, cytochrome P-450, direct demonstration of its function in the reduction of cytochrome P-450 by NADPH was not possible until 1968 when successful reconstitution of the microsomal oxygenase system from solubilized components was first reported by A. Y. H. Lu and M. J. Coon.[57] One of the essential components was NADPH-cytochrome c reductase. The reconstitution experiment has subsequently been repeated using more purified preparations, and it is now clear that only two protein components, cytochrome P-450 and NADPH-cytochrome c reductase, are needed for reconstituting the NADPH-linked oxygenation activity (cf. section 4.1).

Thus we have two distinct types of cytochrome P-450-containing oxygenase systems. One has been found in bacteria and in mitochondria of animal tissues, and consists of three protein components: an NAD(P)H-linked flavoprotein, an iron-sulfur protein, and cytochrome P-450. The other is always found in association with microsomes, and is widely distributed among various eukaryotic organisms ranging from highly developed mammals to unicellular yeasts. The latter type consists of two membrane-bound protein components: an NADPH-linked flavoprotein, which is usually called NADPH-cytochrome c reductase, and cytochrome P-450. Comparison of the properties of the components of these two types of cytochrome P-450 systems shows that the flavoprotein of the former contains only one mole of FAD per mole of enzyme,[54,58,59] whereas that of the latter contains one mole each of FAD and FMN per mole of enzyme.[60] It has been suggested by T. Iyanagi et al.[61] that each of the two flavin molecules in NADPH-cytochrome c reductase has an individual function, and it is tempting to speculate that the FMN-containing portion of the reductase molecule functions in the transfer of electrons from $FADH_2$ to cytochrome P-450 just as the iron-sulfur protein does in bacterial-type cytochrome P-450 systems.

1.5 Elucidation of the Molecular Properties of Cytochrome P-450

Since elucidation of the molecular properties of cytochrome P-450 was largely dependent on the availability of purified homogeneous preparations, attempts were made to solubilize and purify this hemoprotein following its discovery in liver microsomes.[1-3] However, because microsomal cytochrome P-450 is easily converted to P-420, a denatured form, by various treatments including solubilization with detergents,[4] successful isolation of cytochrome P-450 from microsomes or mitochondria had to await the development of some novel contrivance to stabilize it against treatment with detergents. Such an innovation was actually made by Y. Ichikawa and T. Yamano[62] in 1967 when they demonstrated the efficient stabilization of cytochrome P-450 against detergent treatment by polyols including glycerol and by reduced glutathione. Their findings became the most important element in all purification procedures for particle-bound cytochrome P-450's reported in the subsequent years. Another breakthrough in the purification of cytochrome P-450 was the above-mentioned discovery of soluble cytochrome P-450's in bacteria[32,33] which could be purified to homogeneity[36] without much difficulty.

Since recent developments in the purification and properties cytochrome P-450's from various sources will be covered in detail in later chapters, some noteworthy studies carried out when purified preparations of cytochrome P-450 were not yet available will be described briefly here. Due to the fact that the suspensions of microsomes were highly turbid and contained large amounts of cytochrome b_5 in addition to cytochrome P-450, the optical properties of microsomal cytochrome P-450 could initially be studied only by difference spectrophotometry. However, the ambiguous shape of the reduced minus oxidized difference spectrum[4,63] and the unusual shape of the ethyl isocyanide-difference spectrum of reduced cytochrome P-450, which showed two prominent peaks, one at 430 nm and another at 455 nm, in the Soret region,[4,64] were difficult to explain without a knowledge of the actual shapes of the absolute spectra. Thus many investigators became interested in measuring the absolute absorption spectra of cytochrome P-450, and, fortunately, a few important developments made by the mid-1960's rendered observation of the absolute spectra of cytochrome P-450 feasible even before the isolation of this hemoprotein became practicable.

First, the presence of a high concentration of cytochrome P-450 in adrenal cortex mitochondria was discovered by B. W. Harding et al.[21] in 1964. This represented the first case of an occurrence of cytochrome P-450 essentially unaccompanied by cytochrome b_5, and it was possible[23,65] to obtain a cytochrome P-450-containing particle preparation which was virtually free of cytochrome b_5 by sonicating the mitochondrial fraction and fractionating the sonicated particles by centrifugation. Such a preparation was utilized by S. Horie et al.[66] in 1966 for measuring the absolute absorption spectra of cytochrome P-450 for the first time, employing as the reference for spectrophotometry the same particle preparation which had been decolorized with alkali and hydrogen peroxide.

Second, the discovery of an induced increase in cytochrome P-450 in the liver microsomes of drug-treated animals[19] provided further good material for spectral studies on this hemoprotein. Since the content of cytochrome b_5 was not appreciably increased by the drug treatment,[19] it was possible to compensate the absorption due to cytochrome b_5 from the recorded spectra by placing induced and control microsomes in the sample and reference cells, respectively, when the concentrations of the microsomal suspensions were so adjusted as to give the same concentration of cytochrome b_5 in both. This procedure was in fact utilized by some investigators[67,68] in an attempt to observe the absolute absorption spectra of microsomal cytochrome P-450.

Third, the selective removal of cytochrome b_5 from microsomes by protease treatment, which was first described by Omura and Sato in 1963,[69] was also conveniently utilized in spectral studies of cytochrome P-450. Although the treatment caused a significant conversion of cytochrome P-450 to the P-420 form, which remained bound to the microsomes,[5,69] this could mostly be prevented by the presence of glycerol in the digestion medium[70] giving a cytochrome P-450-containing particle preparation suitable for measurement of the absolute spectra of cytochrome P-450.[70] Treatment with a suitable detergent[71] or an organic solvent[72] was also employed for removing cytochrome b_5 from microsomes and obtaining cytochrome P-450-containing particle preparations for spectral studies.[71,72] Owing to these efforts, the absolute absorption spectra of microsomal and mitochondrial cytochrome P-450's were well known before the establishment of procedures for their isolation from the subcellular particles, and the best of the recorded spectra[70,71] were later found to be in good agreement with those for purified cytochrome P-450 preparations (cf. section 3.2).

Another important development in the study of the optical properties of cytochrome P-450 was the discovery of its "substrate-binding spectra" in 1965. The first observation of a substrate-induced spectral shift in cytochrome P-450 was made by S. Narasimhulu[73] when she added 17-hydroxyprogesterone, a substrate of the steroid C21-hydroxylase, to a suspension of adrenal cortex microsomes. The addition of this substrate caused a blue shift in the Soret peak of microsomal cytochrome P-450, giving a difference spectrum with a broad peak at 390 nm and a narrower trough at 420 nm. When C21-hydroxylation of the substrate was initiated by the addition of NADPH, the "steroid-induced" difference spectrum disappeared. A very similar steroid-induced spectral change was also observed with adrenal cortex mitochondria.[74] In this case, cortexone was used as the substrate steroid for 11β-hydroxylation to induce the spectral shift in the mitochondrial cytochrome P-450.

These novel observations with adrenal cortex cytochrome P-450's were soon extended to liver microsomal cytochrome P-450.[75,76] The capability of inducing the spectral change in cytochrome P-450 was not confined to steroid compounds, but a wide variety of substrates including various drugs was found to produce difference spectra when added to suspensions of liver microsomes. Moreover, two distinct types, type I and type II, of substrate-induced difference spectra were recognized.[77] The substrate-induced spectral change was also clearly observed with *Pseudomonas putida* soluble cytochrome P-450.[78] Several lines of evidence confirmed that the spectral change was the result of binding of the substrate with the oxidized form of cytochrome P-450, and that the binding was an essential step in the oxygenation of the substrate (*cf.* Chapter 5). Thus, the finding of the "substrate-binding spectra" of cytochrome P-450 became one of the most important factors in considering the mechanism of the oxygenation reaction catalyzed by cytochrome P-450.

Some of the other physicochemical properties of cytochrome P-450 were also studied before the purification of membrane-bound cytochrome P-450 became feasible. The molecular weight of microsomal cytochrome P-450 had been estimated from electrophoretic patterns of SDS-solubilized microsomes in a polyacrylamide gel. Judging from the effect of phenobarbital, one of the most potent inducers of hepatic microsomal cytochrome P-450, on the protein band intensities, a broad and intense band at a molecular weight of about 50,000 daltons was inferred to represent the band of cytochrome P-450 protein.[79] Another indirect approach was the use of the radiation inactivation of cytochrome P-450 in dried powder of microsomes.[80] This gave a molecular weight of 57,000 daltons for cytochrome P-450. These values were not greatly different from those reported later with purified cytochrome P-450 preparations (*cf.* section 3.1). The oxidation-reduction potential of microsomal cytochrome P-450 was first reported by M. R. Waterman and Mason[81] in 1970. They titrated rabbit liver microsomes with viologen dyes, and determined the extent of reduction of cytochrome P-450 from its characteristic ESR spectra. An apparent mid-potential of -0.34 V at pH 7.0 was calculated for dye-reducible cytochrome P-450.[82] The validity of this very low potential was later confirmed with a highly-purified preparation of this hemoprotein.[83]

Thus, the strenuous efforts of biochemists elucidated many of the basic properties of cytochrome P-450 before pure preparations became available from microsomes or from mitochondria in the 1970's. However, the availability of pure cytochrome P-450's naturally facilitated studies in this direction greatly, and it should soon be possible to determine the detailed chemical properties of microsomal and mitochondrial cytochrome P-450's and to compare them with those of soluble bacterial cytochrome P-450's.

1.6 Multiple Forms of Cytochrome P-450 in Liver Microsomes

Here, a brief review will be given of another interesting problem concerning cytochrome P-450 which has attracted the attention of many biochemists and pharmacologists in recent years; that is, the possible presence of "multiple forms" of cytochrome P-450 in liver microsomes (*cf.* section 4.1).

When the role of cytochrome P-450 in microsome-catalyzed oxygenation reactions of a wide variety of substrates was confirmed by photochemical action spectrum experiments,[11] it was generally believed that only one molecular species of cytochrome P-450 having a broad substrate specificity was participating in all of the reactions since the same action spectrum was obtained for all the reactions examined. However, it was soon found that the mitochondria and microsomes of adrenal cortex also contained cytochrome P-450[21] whose carbon monoxide-difference spectrum was indistinguishable from that of liver microsomes; yet, they catalyze highly specific hydroxylation reactions at particular positions of steroids. The very broad substrate specificity of the cytochrome P-450-dependent oxygenase activity of liver microsomes was rather unusual for a reaction catalyzed by a single enzyme species, and there was actually no strong evidence which could discount the presence of more than one molecular species of cytochrome P-450 in liver microsomes, each of which might catalyze more or less specific oxygenation reactions.

As studies on the drug-induced increase of microsomal drug-oxidizing activities made progress, it became clear[13] that the rates of oxidation of a variety of foreign compounds did not always increase in parallel with one another when animals were treated with different inducers. For example, phenobarbital and 3-methylcholanthrene, when used as inducers, showed very different selective effects on the oxidation of various drugs by the liver microsomes of treated animals.[13] Such observations were provided with a material basis in 1966 when N. E. Sladek and G. J. Mannering[84] reported evidence for the presence of a different species of cytochrome P-450 in liver microsomes from 3-methylcholanthrene-treated rats. The new species of cytochrome P-450 was distinguished from its "normal" counterpart by the different pH dependency of its ethyl isocyanide-difference spectrum, and it was named "cytochrome P_1-450". It was soon found[85] that the carbon monoxide-compound of methylcholanthrene-inducible cytochrome P-450 had an absorption maximum at 448 nm instead of at 450 nm, and the alternate name of "cytochrome P-448" gained wide acceptance.

In the latter half of the 1960's, many papers were published on this subject (cf. section 4.1). The main point of controversy concerned whether the proposed new species of cytochrome P-450 (P_1-450 or P-448) was actually a distinct molecular entity or not. Some investigators[68,86] argued that attachment of the polycyclic hydrocarbon inducer or its metabolites to "normal" cytochrome P-450 could alter the spectral properties of the cytochrome. Some others[87] argued that the drug-induced alteration of the hydrophobic environment (lipids and/or proteins) around the cytochrome P-450 molecules in the membrane could affect the spectral properties and possibly even the substrate specificity of the hemoprotein. However, various lines of evidence for the presence of more than one molecular species of cytochrome P-450 in liver microsomes successively appeared,[88–92] and the solution of the problem had to wait until purification and separation of microsomal cytochrome P-450 became feasible.

The finding of the successful stabilization of cytochrome P-450 by glycerol[62] permitted solubilization and purification experiments on cytochrome P-450 starting from liver microsomes, and, unexpectedly, studies in this direction soon yielded strong evidence for the presence of more than two molecular species of cytochrome P-450 in liver microsomes. SDS-polyacrylamide gel electrophoresis of liver microsomes or of partially purified preparations of microsomal cytochrome P-450 revealed three[93] or four[94] cytochrome P-450 bands. Moreover, these multiple molecular species of cytochrome P-450 could even be detected in the microsomes of normal animals. A report on the separation of microsomal cytochrome P-450 by ion exchange chromatography into three components having different affinities for cyanide[95] provided further supporting evidence for the multiple nature of cytochrome P-450. Partial purification of cytochrome P-450 and "P-448" disclosed different substrate specificities for these two forms.[96–98] A study on the genetic regulation of the microsomal hydroxylase system also supported the presence of a distinct species of cytochrome P-450 active in aryl hydrocarbon hydroxylation.[99]

As more recent purification studies[100–102] on microsomal cytochrome P-450 have further confirmed the presence of more than two distinct molecular species of cytochrome P-450 in liver microsomes, it now appears almost certain that unexpectedly many different forms of cytochrome P-450 are present in individual mammals. Liver microsomes contain three or possibly four forms of the cytochrome. The cytochrome P-450 of kidney microsomes whose carbon monoxide-difference spectrum shows an absorption peak at 454 nm instead of at 450 nm,[103] is certainly different from liver microsomal cytochrome P-450 in its substrate specificity[103] as well as its spectral properties. The microsomes of adrenal cortex contain another different form of cytochrome P-450,[73] and mitochondria from the same tissue contain at least two molecular species of the cytochrome,[104] etc. Thus, commencing with the studies on the multiple forms of cytochrome P-450 in liver microsomes, we have now arrived at the important conclusion that the name "cytochrome P-450" embraces a group of many hemoproteins having very similar molecular configurations around a common prosthetic group, protohemin. Since we do not yet have much accurate knowledge on the primary amino acid sequence and conformation of these many molecular species of cytochrome P-450, it is not clear at present whether they are intimately related structurally with one another or not, although their molecular weights are all close to 50,000 daltons. Further purification studies on these various forms of cytochrome P-450 from different animal tissues, and detailed chemical studies on the purified preparations obtained, will undoubtedly solve this problem.

1.7 Present Problems and Future Prospects

As the history of cytochrome P-450 is still short, many unanswered problems remain concerning this hemoprotein. The immediate questions to be studied are undoubtedly the chemical nature of the cytochrome P-450 molecules, the configuration of the active site of cytochrome P-450, and the mechanism of activation of the oxygen molecule at the active site, since pure preparations of cytochrome P-450 have just become available from microsomes and mitochondria. Elucidation of the molecular structure of cytochrome P-450 will also enable us to investigate how the substrate specificity of cytochrome P-450-catalyzed oxygenation reactions is determined. Rapid progress in these fields can be expected in the next few years, and recent developments in each of them are covered in later chapters of this book.

When the molecular properties of cytochrome P-450 and the constitution of cytochrome P-450-containing oxygenase systems become better known, our attention should be focused on the mechanism of regulation of the enzyme systems. This means that we will again return to intact microsomes and mitochondria, or even to intact cells, from the present studies on purified enzymes and reconstituted systems. As all components of the microsomal cytochrome P-450 system are membrane-bound, information about their topographical distribution in the membrane is highly desirable for understanding the mechanism of regulation of this enzyme system. It is very likely that alterations in the distribution of cytochrome P-450 and its reductase molecules in the membrane affect the activity of cytochrome P-450-catalyzed oxygenation reactions. Even in the case of the mitochondrial cytochrome P-450 system, where only cytochrome P-450 is bound to the membrane while its reductase system consisting of a flavoprotein and an iron-sulfur protein is soluble in the matrix compartment, the extent of association, if there is any, among the cytochrome P-450 molecules and the molecules of the reductase system will affect the overall activity of the enzyme system.

The activity of cytochrome P-450-containing oxygenase systems may also be regulated by alterations in the amounts of all or some of the components of the enzyme systems. The drug-induced increase of microsomal cytochrome P-450 (or P-448) and its reductase, NADPH-cytochrome c reductase, is a well-known example of regulation of this type in which the rate of synthesis and the rate of degradation of the reductase are both affected by the drug.[105,106] A similar mechanism may also operate under physiological conditions, and it seems likely that all cytochrome P-450 enzyme systems are inducible enzyme systems whose activities are maintained *in vivo* through continuous induction by some natural inducers, although as yet we have very little knowledge about them. One remarkable example is the pronounced circadian rhythm of the cholesterol 7α-hydroxylase activity of rat liver microsomes[107] which has recently been shown to involve a rapid turnover of a certain component of the enzyme system, possibly a particular species of cytochrome P-450. The circadian rhythm of this hydroxylase activity is somehow controlled by the feeding time of the animals. Studies in this direction will also make progress in the coming years, and it can be expected that a better understanding of how the synthesis and degradation of cytochrome P-450 and its reductase is regulated *in vivo* will be obtained.

Concerning the physiological significance of cytochrome P-450 enzyme systems, one recent topic which has attracted broad attention among biochemists is the possible involvement of cytochrome P-450-containing oxygenase systems in the metabolic activation of certain carcinogenic compounds, particularly polycyclic aromatic hydrocarbons, causing their covalent binding to cellular macromolecules (*cf.* the review articles, ref. 108 and 109). The introduction of a hydroxyl group into the aromatic ring of a polycyclic hydrocarbon by a cytochrome P-450-catalyzed oxygenation reaction is not a one-step reaction, but an epoxide (arene oxide) is formed as an intermediate[110,111] which then undergoes non-enzymatic isomerization to a phenol, enzymatic hydration to a dihydrodiol, or enzyme-catalyzed conjugation with glutathione. However, the arene oxide can also react with proteins or nucleic acids and bind with them covalently.[108,109] The significance of this reaction in chemical carcinogenesis is without doubt. Thus, the microsomal cytochrome P-450 enzyme system, which we usually consider as functioning in the "detoxification" of foreign compounds, can also be active in transforming relatively inert foreign compounds into toxic and carcinogenic metabolites. The metabolism of xenobiotics in the animal body is certainly much more complicated than previously thought, and it will be necessary to study various aspects of cytochrome P-450-catalyzed reactions in order to obtain a clearer picture of the physiological significance of cytochrome P-450.

The wide occurrence of cytochrome P-450 among various forms of life from higher animals to bacteria, suggests that the evolutionary aspects of this hemoprotein may also represent an interesting and important subject for study. Whereas the microsomal cytochrome P-450-containing oxygenase system of eukaryotic cells is very tightly integrated in the membrane structure, the corresponding enzyme system in bacterial cells consists of soluble enzyme proteins present in the cytoplasm. The mitochondrial cytochrome P-450 enzyme system, in which only cytochrome P-450 is membrane-bound while its reductase system is soluble in the matrix space, appears to be an intermediate form between the two. If cytochrome P-450 has in fact evolved from a primitive soluble form to diversified membrane-bound molecular species, this hemoprotein would represent a very interesting example of membrane proteins from which it might be possible to explore how such enzyme proteins have acquired a hydrophobic property during their molecular evolution to become incorporated into biomembranes. Protein chemical studies on various molecular species of cytochrome P-450 obtained from diversified organisms, which are now just beginning, will undoubtedly permit us later to speculate on such biological problems.

References

1. M. Klingenberg, *Arch. Biochem. Biophys.*, **75**, 376 (1958).
2. D. Garfinkel, *ibid.*, **77**, 493 (1958).

3. T. Omura and R. Sato, *J. Biol. Chem.*, **237**, 1375 (1962).
4. T. Omura and R. Sato, *ibid.*, **239**, 2370 (1964).
5. T. Omura and R. Sato, *ibid.*, **239**, 2379 (1964).
6. D. Ryan, A. Y. H. Lu, J. Kawalek, S. B. West and W. Levin, *Biochem. Biophys. Res. Commun.*, **64**, 1134 (1975).
7. Y. Hashimoto, T. Yamano and H. S. Mason, *J. Biol. Chem.*, **237**, 3843 (1962).
8. Y. Ichikawa, B. Hagihara and T. Yamano, *Arch. Biochem. Biophys.*, **120**, 204 (1967).
9. Y. Ichikawa and T. Yamano, *ibid.*, **121**, 742 (1967).
10. R. W. Estabrook, D. Y. Cooper and O. Rosenthal, *Biochem. Z.*, **338**, 741 (1963).
11. D. Y. Cooper, S. S. Levin, S. Narasimhulu, O. Rosenthal and R. W. Estabrook, *Science*, **147**, 400 (1965).
12. D. Y. Cooper, *Life Sci.*, **13**, 1151 (1973).
13. A. H. Conney, *Pharmacol. Rev.*, **19**, 317 (1967).
14. G. C. Mueller and J. A. Miller, *J. Biol. Chem.*, **180**, 1125 (1949).
15. H. Remmer, *Arch. Exptl. Pathol. Pharmakol.*, **237**, 296 (1959).
16. R. Kato, *Experientia*, **16**, 9 (1960).
17. A. H. Conney, C. Davidson, R. Gastel and J. J. Burns, *J. Pharmacol. Exptl. Ther.*, **130**, 1 (1960).
18. L. Ernster and S. Orrenius, *Fed. Proc.*, **24**, 1190 (1965).
19. S. Orrenius, J. L. E. Ericsson and L. Ernster, *J. Cell Biol.*, **25**, 627 (1965).
20. R. Sato and T. Omura, *Oxidases and Related Redox Systems* (ed. T. E. King, H. S. Mason and M. Morrison), vol. 2, p. 861, John Wiley (1965).
21. B. W. Harding, S. H. Wong and D. H. Nelson, *Biochim. Biophys. Acta*, **92**, 415 (1964).
22. T. Omura, R. Sato, D. Y. Cooper, O. Rosenthal and R. W. Estabrook, *Fed. Proc.*, **24**, 1181 (1965).
23. T. Omura, E. Sanders, R. W. Estabrook, D. Y. Cooper and O. Rosenthal, *Arch. Biochem. Biophys.*, **117**, 660 (1966).
24. E. R. Simpson and G. S. Boyd, *Biochem. Biophys. Res. Commun.*, **24**, 10 (1966).
25. E. R. Simpson and G. S. Boyd, *Eur. J. Biochem.*, **2**, 275 (1967).
26. T. Yohro and S. Horie, *J. Biochem. (Tokyo)*, **61**, 515 (1967).
27. R. A. Meigs and K. J. Ryan, *Biochim. Biophys. Acta*, **165**, 476 (1968).
28. S. Taniguchi, N. Hoshita and K. Okuda, *Eur. J. Biochem.*, **40**, 607 (1973).
29. H. L. Henry and A. W. Norman, *J. Biol. Chem.*, **249**, 7529 (1974).
30. J. G. Ghazarian, C. R. Jefcoate, J. C. Knutson, W. H. Orme-Johnson and H. F. DeLuca, *ibid.*, **249**, 3026 (1974).
31. A. Lindenmayer and L. Smith, *Biochim. Biophys. Acta*, **93**, 445 (1964).
32. C. A. Appleby, *ibid.*, **147**, 399 (1967).
33. M. Katagiri, B. N. Ganguli and I. C. Gunsalus, *J. Biol. Chem.*, **243**, 3543 (1968).
34. G. Cardini and P. Jurtshuk, *ibid.*, **243**, 6070 (1968).
35. P. F. Hollenberg and L. P. Hager, *ibid.*, **248**, 2630 (1973).
36. K. Dus, M. Katagiri, C. A. Yu, D. L. Erbes and I. C. Gunsalus, *Biochem. Biophys. Res. Commun.*, **40**, 1423 (1970).
37. C. A. Yu and I. C. Gunsalus, *ibid.*, **40**, 1431 (1970).
38. P. J. Murphy and C. A. West, *Arch. Biochem. Biophys.*, **133**, 395 (1970).
39. J. R. M. Potts, R. Weklych and E. E. Conn, *J. Biol. Chem.*, **249**, 5019 (1974).
40. T. D. Meehan and C. J. Coscia, *Biochem. Biophys. Res. Commun.*, **53**, 1043 (1973).
41. J. W. Ray, *Biochem. Pharmacol.*, **16**, 99 (1967).
42. S. E. Lewis, *Nature*, **215**, 1408 (1967).
43. A. S. Perry, *Life Sci.*, **9**, 335 (1970).
44. A. Morello, W. Bleeker and M. Agosin, *Biochem. J.*, **124**, 199 (1971).
45. J. Capdevila, A. S. Perry, A. Morello and M. Agosin, *Biochemistry*, **12**, 1445 (1973).
46. T. M. Chan, J. W. Gillet and L. C. Terriere, *Comp. Biochem. Physiol.*, **20**, 731 (1967).
47. J. H. Dewaide, *Comp. Gen. Physiol.*, **1**, 375 (1970).

48. J. H. Dewaide and P. T. Henderson, *Comp. Biochem. Physiol.*, **32**, 489 (1970).
49. T. Ahokas, O. Pelkonen and N. T. Karki, *Biochem. Biophys. Res. Commun.*, **63**, 635 (1975).
50. K. Suzuki and T. Kimura, *ibid.*, **19**, 340 (1965).
51. T. Kimura and K. Suzuki, *ibid.*, **20**, 373 (1965).
52. Y. Nakamura, H. Otsuka and B. Tamaoki, *Biochim. Biophys. Acta*, **122**, 34 (1966).
53. D. W. Cushman, R. L. Tsai and I. C. Gunsalus, *Biochem. Biophys. Res. Commun.*, **26**, 577 (1967).
54. R. L. Tsai, I. C. Gunsalus and K. Dus, *ibid.*, **45**, 1300 (1971).
55. T. Kimura and H. Ohno, *J. Biochem. (Tokyo)*, **63**, 716 (1968).
56. A. H. Philips and R. G. Langdon, *J. Biol. Chem.*, **237**, 2652 (1962).
57. A. Y. H. Lu and M. J. Coon, *ibid.*, **243**, 1331 (1968).
58. K. Suhara, Y. Ikeda, S. Takemori and M. Katagiri, *FEBS Lett.*, **28**, 45 (1972).
59. J. W. Chu and T. Kimura, *J. Biol. Chem.*, **248**, 2089 (1973).
60. T. Iyanagi and H. S. Mason, *Biochemistry*, **12**, 2297 (1973).
61. T. Iyanagi, N. Makino and H. S. Mason, *ibid.*, **13**, 1701 (1974).
62. Y. Ichikawa and T. Yamano, *Biochim. Biophys. Acta*, **131**, 490 (1967).
63. H. Nishibayashi and R. Sato, *J. Biochem. (Tokyo)*, **61**, 491 (1967).
64. Y. Imai and R. Sato, *Biochem. Biophys. Res. Commun.*, **23**, 5 (1966).
65. T. Kinoshita, S. Horie, N. Shimazono and T. Yohro, *J. Biochem. (Tokyo)*, **60**, 391 (1966).
66. S. Horie, T. Kinoshita and N. Shimazono, *ibid.*, **60**, 660 (1966).
67. T. Kinoshita and S. Horie, *ibid.*, **61**, 26 (1967).
68. A. G. Hildebrandt, H. Remmer and R. W. Estabrook, *Biochem. Biophys. Res. Commun.*, **30**, 607 (1968).
69. T. Omura and R. Sato, *Biochim. Biophys. Acta*, **71**, 224 (1963).
70. H. Nishibayashi and R. Sato, *J. Biochem. (Tokyo)*, **63**, 766 (1968).
71. Y. Miyake, J. L. Gaylor and H. S. Mason, *J. Biol. Chem.*, **243**, 5788 (1968).
72. D. H. McLennan, A. Tzagoloff and D. G. McConnell, *Biochim. Biophys. Acta*, **131**, 59 (1967).
73. S. Narasimhulu, D. Y. Cooper and O. Rosenthal, *Life Sci.*, **4**, 2101 (1965).
74. D. Y. Cooper, S. Narasimhulu, A. Slade, W. Raich, O. Foroff and O. Rosenthal, *ibid.*, **4**, 2109 (1965).
75. H. Remmer, J. B. Schenkman, R. W. Estabrook, H. Sasame, J. R. Gillette, D. Y. Cooper, S. Narasimhulu and O. Rosenthal, *Mol. Pharmacol.*, **2**, 187 (1966).
76. Y. Imai and R. Sato, *Biochem. Biophys. Res. Commun.*, **22**, 620 (1966).
77. J. B. Schenkman, H. Remmer and R. W. Estabrook, *Mol. Pharmacol.*, **3**, 113 (1967).
78. I. C. Gunsalus, *Z. Physiol. Chem.*, **349**, 1610 (1968).
79. P. J. Dehlinger and R. T. Schimke, *J. Biol. Chem.*, **247**, 1257 (1972).
80. M. Shikita and F. Hatano-Sato, *FEBS Lett.*, **36**, 187 (1973).
81. M. R. Waterman and H. S. Mason, *Biochem. Biophys. Res. Commun.*, **39**, 450 (1970).
82. M. R. Waterman and H. S. Mason, *Arch. Biochem. Biophys.*, **150**, 57 (1972).
83. F. P. Guengerich, D. P. Ballow and M. J. Coon, *J. Biol. Chem.*, **250**, 7405 (1975).
84. N. E. Sladek and G. J. Mannering, *Biochem. Biophys. Res. Commun.*, **24**, 668 (1966).
85. A. P. Alvares, G. Schilling, W. Levin and R. Kuntzman, *ibid.*, **29**, 521 (1967).
86. J. B. Schenkman, H. Greim, M. Zauge and H. Remmer, *Biochim. Biophys. Acta*, **171**, 23 (1969).
87. Y. Imai and P. Siekevitz, *Arch. Biochem. Biophys.*, **144**, 143 (1971).
88. A. P. Alvares, G. Schilling, W. Levin and R. Kuntzman, *J. Pharmacol. Exptl. Ther.*, **163**, 417 (1968).
89. K. Einarsson and G. Johansson, *Eur. J. Biochem.*, **6**, 293 (1968).
90. C. R. E. Jefcoate and J. L. Gaylor, *Biochemistry*, **8**, 3464 (1969).
91. W. Levin and R. Kuntzman, *J. Biol. Chem.*, **244**, 3671 (1969).

92. F. J. Wiebel, J. C. Leutz, L. Diamond and H. V. Gelboin, *Arch. Biochem. Biophys.*, **144**, 78 (1971).
93. A. F. Welton and S. D. Aust, *Biochem. Biophys. Res. Commun.*, **56**, 898 (1974).
94. A. P. Alvares and P. Siekevitz, *ibid.*, **54**, 923 (1973).
95. K. Comai and J. L. Gaylor, *J. Biol. Chem.*, **248**, 4947 (1973).
96. A. Y. H. Lu and W. Levin, *Biochem. Biophys. Res. Commun.*, **46**, 1334 (1972).
97. A. Y. H. Lu, R. Kuntzman, S. West, M. Jacobson and A. H. Conney, *J. Biol. Chem.*, **247**, 1727 (1972 .
98. D. W. Nebert, J. K. Heidema, H. W. Strobel and M. J. Coon, *ibid.*, **248**, 7631 (1973).
99. J. E. Gielen, F. M. Goujon and D. W. Nebert, *ibid.*, **247**, 1125 (1972).
100. J. Capdevila, N. Ahmad and M. Agosin, *ibid.*, **250**, 1048 (1975).
101. D. A. Haugen, T. A. van der Hoeven and M. J. Coon, *ibid.*, **250**, 3567 (1975).
102. A. F. Welton, F. O. O'Neal, L. C. Chaney and S. D. Aust, *ibid.*, **250**, 5631 (1975).
103. S. Jakobsson, H. Thor and S. Orrenius, *Biochem. Biophys. Res. Commun.*, **39**, 1073 (1970).
104. S. Takemori, H. Sato, T. Gomi, K. Suhara and M. Katagiri, *ibid.*, **67**, 1151 (1975).
105. H. Jick and L. Shuster, *J. Biol. Chem.*, **241**, 5366 (1966).
106. Y. Kuriyama, T. Omura, P. Siekevitz and G. E. Palade, *ibid.*, **244**, 2017 (1969).
107. J. Gielen, J. van Cantfort, B. Robaye and J. Renson, *Eur. J. Biochem.*, **55**, 41 (1975).
108. P. L. Grover and P. Sims, *Advan. Cancer. Res.*, **20**, 166 (1974).
109. D. M. Jerina and J. W. Daly, *Science*, **185**, 573 (1974).
110. D. M. Jerina, J. W. Daly, B. Witkop, P. Zaltman-Nirenberg and S. Udenfriend, *Biochemistry*, **9**, 147 (1970).
111. J. K. Selkirk, E. Huberman and C. Heidelberger, *Biochem. Biophys. Res. Commun.*, **43**, 1010 (1971).

CHAPTER 2

Distribution and Physiological Functions

2.1. Introduction
2.2. Distribution of Cytochrome P-450
2.3. Catalytic Activities of Cytochrome P-450
2.4. Metabolic Roles of Cytochrome P-450

2.1 INTRODUCTION

Cytochrome P-450 is the collective name for a distinct group of protoheme-contaning proteins which show a Soret absorption band at around 450 nm (447 to 454 nm) in the CO–difference spectrum of dithionite-reduced samples. This and certain other spectral properties of these hemoproteins are quite atypical and suggest the possession of a unique electronic structure in the vicinity of the protoheme prosthetic group. Recent studies have provided evidence that the anomalous spectral properties are a reflection of the heme environment in which a thiolate or mercaptide anion ($-S^-$), probably derived from a cysteinyl residue, is coordinated to the heme iron.[1–3] The purpose of this chapter is to summarize our present knowledge on the distribution and physiological functions of cytochrome P-450.

2.2 DISTRIBUTION OF CYTOCHROME P-450

Cytochrome P-450 was first described in mammalian liver microsomes[4–8] and was originally thought to be a curiosity occurring only in restricted biological systems. However, subsequent studies have revea-

* Ryo SATO, Institute for Protein Research, Osaka University, Suita, Osaka 565, Japan

led that hemoproteins with similar spectral properties are distributed very widely in nature: they can be detected in almost all forms of life.

In mammals, cytochrome P-450's are found at varying concentrations in microsomes (endoplasmic reticulum) of the liver,[4-8] kidney,[9] small intestine,[10] lung,[11] adrenal cortex,[12] skin,[13] testis,[14] placenta,[15] and several other tissues. However, microsomes from other organs such as the brain, muscle, and thyroid gland appear to be devoid of this type of hemoprotein. Cytochrome P-450's have also been detected in microsomes from various tissues, notably the liver, of non-mammalian vertebrates, i.e. birds, reptiles, amphibians, and fish,[16,17]. In isolated hepatic microsomal vesicles, cytochrome P-450 is inhibited by exogenously added antibodies against the hemoprotein,[18] suggesting that the antigenic site and probably also the catalytic site of the hemoprotein are exposed to the outer surface of the microsomal membrane.

Mitochondria of mammalian endocrine glands such as the adrenal cortex[19], testis[20], and corpus luteum[21], which synthesize and excrete steroid hormones, invariably contain cytochrome P-450's. However, the mitochondrial localization of this group of hemoproteins is not restricted to steroid hormone-producing glands. Recently, the occurrence of cytochrome P-450's in mitochondria of chicken kidney[22,23] and rat liver [24,25] has been confirmed, although the concentrations in such mitochondria are rather low. It is very likely, therefore, that cytochrome P-450's also occur in other types of mitochondria. At least in the mitochondria of bovine adrenal cortex[26] and rat liver,[25] there is evidence that the hemoproteins are associated with the inner mitochondrial membrane.

Besides microsomes and mitochondria, the nuclear envelope of rat liver has been reported to contain a cytochrome P-450.[27] This does not appear to be due to microsomal contamination, since the cytochrome, unlike its microsomal counterpart, cannot be induced by phenobarbital treatment of the animals, although it is normally inducible by 3-methylcholanthrene administration.[27] A few reports have indicated the presence of a cytochrome P-450 in the rat liver Golgi apparatus,[28] although the validity of this finding requires checking with Golgi preparations of higher purity. Although no detailed studies have as yet appeared, other cellular organelles such as the plasma membrane, lysosomes, and peroxisomes appear to lack cytochrome P-450's.

Studies on the distribution of cytochrome P-450's in invertebrates are not yet extensive, but in connection with the work on insecticide metabolism it has been shown that microsomal fractions from whole insects such as the housefly and *Drosophila* contain significant amounts of this type of hemoprotein.[29-31]

In higher plants, cytochrome P-450's have been discovered during attempts to elucidate the pathways involved in specialized metabolisms such as the biosynthesis of phytohormones, lignins, and alkaloids. To date, the occurrence of cytochrome P-450's has been confirmed in such sources as sorghum seedlings,[32] cauliflower buds,[33] mungbean hypocotyls,[33] castor bean endosperm,[34] *Echinocystis macrocarpa* endosperm[35], and *Catharanthus roseus* (*Vinca roseo*) seedlings.[36] In all of these plant sources the hemoproteins were reported to be localized in the microsomal fractions; however, a recent study has indicated that in *C. roseus* seedlings the cytochrome P-450 is largely associated with the membrane of provacuoles which are believed to have differentiated from the endoplasmic reticulum.[37]

A cytochrome P-450 occurs in the yeast *Saccharomyces cerevisiae*,[38] particularly in cells grown semianaerobically,[39] and it is also localized in the microsomal fraction.[39] Microsomes from several molds such as *Penicillium patulum*[40] and *Claviceps purpurea*[41] have been shown to contain this class of hemoproteins. Eukaryotic microorganisms such as *Candida tropicalis* grown on tetradecane as carbon source[42] and *Streptomyces erythreus*[43] have, however, been reported to possess cytochrome P-450's that are apparently not membrane-bound and can be recovered in the soluble fraction. The cytochrome P-450 of *C. tropicalis*, though apparently soluble, requires phospholipids for its activity,[42] suggesting the possibility that it is actually associated with membranes *in vivo*.

Localization in the soluble cytoplasmic fraction seems to be a general feature of bacterial cytochrome P-450's. All cytochromes of this type so far identified in bacteria, e.g. in *Pseudomonas putida* grown on camphor as carbon source,[44] *Corynebacterium* species,[45] bacteroids of *Rhizobium japonicum*,[46] and *Bacillus megaterium*,[47] have been isolated in soluble form. No report is yet available on the distribution of cytochrome P-450 in blue-green algae.

The fact that cytochrome P-450's can be found even in prokaryotic organisms suggests that the unique electronic structure around the heme, a characteristic feature of this type of hemoproteins, is not a recent addition in the evolutionary process but has been utilized by life since ancient times. However, all the bacteria so far reported to contain cytochrome P-450's are aerobic; no strictly anaerobic bacteria have been shown to synthesize this type of hemoproteins. It might, therefore, be suggested that the evolutionary emergence of hemoproteins of the cytochrome P-450 type was an event which occurred after primitive organisms had adapted to utilize atmospheric oxygen. Another interesting fact from the viewpoint of comparative biochemistry is that soluble cytochrome P-450's are almost exclusively found in prokaryotic organisms. Higher forms of life, with only a few dubious exceptions, always contain the hemoproteins in a membrane-bound state. The significance of this fact is at present unclear, but a conversion from the soluble to membrane-bound state during evolution has also been recognized in several other enzymes and enzyme systems.

Direct and indirect evidence accumulated in recent years has introduced a complication into the distribution of cytochrome P-450's. The evidence indicates that multiple forms of the hemoprotein differing from one another in both molecular and catalytic properties can occur in the same individual sources. Thus, recent purification studies have established that several (at least two and very probably more) different species of cytochrome P-450 are present in mammalian liver microsomes, and pretreatment of the animals with drugs such as phenobarbital, 3-methylcholanthrene, and polychlorinated biphenyls appears to induce rather specifically the synthesis of one or two of the multiple forms.[48-53] Two clearly different molecular species of cytochrome P-450, one responsible for cholesterol side-chain cleavage and the other for steroid 11β-hydroxylation, have been purified to homogeneity from bovine adrenocortical mitochondria.[54,55] None of these mitochondrial cytochrome P-450's seems to be identical with those occurring in liver microsomes. Thus, each mammalian organism must have many structural genes coding for the different species of cytochrome P-450. The multiplicity of cytochrome P-450 has also been demonstated in such systems as rat testis microsomes,[14] housefly microsomes,[31] and *R. japonicum* bacteroids.[56]

2.3 Catalytic Activities of Cytochrome P-450

Cytochrome P-450's, unlike many other cytochromes, are not mere electron carriers but are enzymes catalyzing several types of redox reactions. It appears that the most fundamental function of this group of hemoproteins is to catalyze the monooxygenase (hydroxylation) reactions of a variety of substrates by utilizing both molecular oxygen and reducing equivalents (electrons) derived from either NADPH or NADH. These reactions can be expressed by the following equation.

$$S + NAD(P)H + H^+ + O_2 \longrightarrow SO + NAD(P)^+ + H_2O, \qquad (2.1)$$

where S is the substrate.

In order for this type of reaction to take place, however, cooperation of a mechanism by which electrons can be transferred from NAD

(P)H to cytochrome P-450 is required, since the hemoprotein itself is incapable of reacting directly with reduced pyridine nucleotides. Two different types of mechanism are apparently operative for this purpose. Thus, in camphor-grown *P. putida*, a flavoprotein containing FAD as prosthetic group (putidaredoxin reductase) and an iron-sulfur protein of ferredoxin type(putidaredoxin) participate in the electron transfer from NADH to cytochrome P-450.[44] This pathway can be expressed as follows.

$$\text{NADH} \longrightarrow \text{flavoprotein (FAD)} \longrightarrow \text{iron-sulfur protein} \longrightarrow \text{cytochrome P-450} \quad (2.2)$$

A closely analogous system has also been shown to be functional in *B. megaterium*,[47] although in this system, NADPH, rather than NADH, serves as the electron donor. In these bacterial systems, all components are present in the soluble fraction of the cells. Mammalian and avian mitochondria in which cytochrome P-450's occur also possess essentially the same type of electron transfer mechanism. For instance, the reduction of cytochrome P-450 in bovine adrenocortical mitochondria is effected by a system consisting of an NADPH-specific, FAD-containing flavoprotein (adrenodoxin reductase) and an iron-sulfur protein (adrenodoxin).[57,58] In contrast to the bacterial systems, the components of this mitochondrial system are membrane-bound, although both adrenodoxin reductase and adrenodoxin can be readily solubilized from the membrane by mild treatment. The cytochrome P-450-reducing electron transfer systems in the mitochondria of chicken kidney[23] and rat liver[25] belong to the same category. The occurrence of similar electron transfer mechanisms in both bacteria and mitochondria is of special interest in relation to the symbiotic hypothesis for the origin of mitochondria,[59] in which it is assumed that the mitochondria of eukaryotic cells have evolved from prokaryotic organisms which were originally in a symbiotic relationship with the host cells.

The other main type of electron transfer system involved in the reduction of cytochrome P-450's is that found in liver microsomes, where NADPH is the principal electron donor and the reducing equivalents are transferred to the cytochrome via a flavoprotein without intervention of an iron-sulfur protein.[60,61] The flavoprotein, which has high NADPH-cytochrome *c* reductase activity, has been shown to contain both FAD and FMN as prosthetic groups.[62] This pathway can, therefore, be expressed as follows.

$$\text{NADPH} \longrightarrow \text{flavoprotein (FAD and FMN)} \longrightarrow \text{cytochrome P-450} \quad (2.3)$$

In this system, both the flavoprotein and the cytochrome are tightly attached to the membrane and their interaction requires the presence of phosphatidylcholine.[63] Evidence has been presented that the same type of mechanism is also operative in microsomes of the kidney[64] and of *S. cerevisiae*,[65] It appears likely that this mechanism is universally functional in almost all microsomal cytochrome P-450-containing systems of different origin.

One complication in the hepatic microsomal system is that NADH, although much less effective than NADPH, can also act as an electron donor for cytochrome P-450-catalyzed monooxygenase reactions.[66] Reconstitution as well as immunochemical studies have indicated that the electron transfer in this case is mediated by an FAD-containing flavoprotein (NADH-cytochrome b_5 reductase) and a b-type cytochrome (cytochrome b_5) as shown in the following scheme.[67]

$$\text{NADH} \xrightarrow{\text{flavoprotein (FAD)}} \text{cytochrome } b_5 \longrightarrow \text{cytochrome P-450} \qquad (2.4)$$

It has further been reported that NADH exerts a synergistic effect on NADPH-dependent monooxygenase reactions catalyzed by liver microsomes.[68-70] Although much remains unclear about this synergism, evidence from immunochemical[71] and reconstitution[72] experiments suggests that the electron transfer pathway in Eq. (2.4) is somehow involved in this phenomenon.

In cytochrome P-450-catalyzed monooxygenase reactions, one atom of oxygen is incorporated into the substrate (S) to form an oxygenated product (SO), as shown in Eq. (2.1). In a number of instances, however, SO is chemically unstable and undergoes spontaneous decomposition. For example, introduction of an oxygen atom into the methyl moiety of an N-monomethyl arylamine leads to the formation of a labile intermediate which rapidly decomposes to the arylamine and formaldehyde.

$$\text{R-NHCH}_3 \xrightarrow{[O]} (\text{R-NHCH}_2\text{OH}) \xrightarrow{\text{spontaneous}} \text{R-NH}_2 + \text{HCHO} \qquad (2.5)$$

As mentioned above, catalysis of monooxygenase reactions in the presence of both molecular oxygen and reduced pyridine nucleotides appears to represent the most important function of hemoproteins of the cytochrome P-450 type, although several other auxiliary functions have been reported. For example, it has been shown that in liver microsomes both NADPH and NADH and be anaerobically oxidized by organic hydroperoxides such as cumene hydroperoxide, as well as by hydrogen peroxide, and that these peroxidatic reactions are catalyzed by the cytochrome P-450-containing electron transfer mechanisms discussed above.[73,74] More recently, hepatic microsomal cytochrome P-450 has been shown to catalyze the oxidations of a variety of drugs in the absence of NAD(P)H and molecular oxygen provided that hydrogen peroxide, organic hydroperoxides, NaIO$_4$, etc. are supplied.[75,76] It is thought that this type of reaction represents certain partial reaction steps of the monooxygenase reaction which occurs in the presence of both molecular oxygen and reduced pyridine nucleotides. The reducing equivalents transferred from NAD(P)H to liver microsomal cytochrome P-450 can also be utilized under anaerobic conditions to reduce organic nitro and azo groups.[77,78] They can further be used for the reduction of tertiary amine N-oxides.[79] However, it remains to be determined whether or not these auxiliary functions are of significance in relation to the metabolic roles of cytochrome P-450's.

2.4 Metabolic Roles of Cytochrome P-450

The first report dealing with the metabolic function of hemoproteins of the cytochrome P-450 type was published by Estabrook et al.,[12] who demonstrated by a photochemical action spectrum technique that a cytochrome P-450 in the microsomal fraction of bovine adrenal cortex is functional in steroid 21-hydroxylation. This important finding was followed by the discovery that liver microsomal cytochrome P-450 is involved in the NADPH-dependent oxidations of various drugs.[80] Subsequently, a number of metabolic roles have been established or proposed for cytochrome P-450's of different origins and the metabolic importance of this family of hemoproteins is now well recognized.

As can be imagined from the fact that the first report on the metabolic function of cytochrome P-450 concerned steroid 21-hydroxylation, many cytochrome P-450's, especially those occurring in mammalian tissues and yeast, have been shown to be involved in various facets of steroid metabolism. In microsomes of both mammalian liver[81] and *S. cerevisiae*,[82] there is evidence that one or more cytochrome P-450's participate in the biosynthesis of cholesterol and ergosterol, respectively, from acetate or mevalonate. Although the step(s) in which the hemoprotein(s) are involved are not yet clear, it is likely that cytochrome P-450 is required at least for the oxidative removal of three methyl groups at positions 4 and 14 of lanosterol. Several important steps in the conversion of cholesterol to bile acids in liver microsomes (notably cholesterol 7α-hydroxylation[83]) and in liver mitochondria (26-hydroxylation of 5β-cholestane-3α,7α,12α-triol[24,25]) are also catalyzed by cytochrome P-450-containing monooxygenase systems.

Cytochrome P-450's are further involved in almost all the metabolic pathways leading to the synthesis of steroid hormones in various endocrine glands. The first step common to all these sequences is the side-chain cleavage of cholesterol to yield pregnenolone. At least in adrenocortical mitochondria, this step has been shown to be catalyzed by a rather specific species of cytochrome P-450.[84] As mentioned above, adrenocortical mitochondria contain another form of cytochrome P-450 which is responsible for the second step of corticosteroid biosynthesis, i.e. 11β-hydroxylation.[55] Numerous reports indicate that cytochrome P-450's are functional in the formation of various androgens and estrogens from pregnenolone in various reproductive glands. It should be noted that in placental microsomes a cytochrome P-450 catalyzes the conversion of androstenedione to estrone, a reaction which involves 19-demethylation and aromatization of ring A of the steroid nucleus.[85,86] Further, many steroid hormones appear to be inactivated by liver microsomes in reactions involving cytochrome P-450.

Recent studies have revealed that vitamin D_3 (cholecalciferol) as such is not physiologically active and has to be metabolically activated in the body in order to fulfill its biological function. This activation involves 25-hydroxylation in liver microsomes and subsequent 1-hydroxylation in kidney mitochondria, and it is now established that cytochrome P-450-linked monooxygenase systems are involved in these two hydroxylation reactions.[22,23,87] Another function of cytochrome P-450's in mammalian tissues is to catalyze the ω-hydroxylation of medium-chain saturated fatty acids; this activity has been detected in microsomes of the liver[88] and kidney.[89]

In the metabolism of higher plants, cytochrome P-450's have been shown to be involved in at least three processes. First, in wheat germ microsomes a cytochrome P-450 is responsible for the hydroxylation of kaurene to kaurenol, an important step in the biosynthesis of the phytohormone, gibberellin.[90] Second, the involvement of a cytochrome P-450 in the 4-hydroxylation of cinnamic acid in microsomes of sorghum seedlings has been reported.[32] It should be emphasized in this connection that 4-hydroxycinnamic acid is a precursor of lignins, tannins and flavonoids, which are produced in enormous quantities by many higher plants. Third, in *C. roseus* seedlings a microsomal cytochrome P-450 catalyzes the 10-hydroxylation of monoterpene alcohols, geraniol and nerol;[36] this metabolic sequence represents the beginning of indole alkaloid biosynthesis. Ergot alkaloid biosynthesis by the fungus *Claviceps purpurea* has also been reported to be a cytochrome P-450-dependent process.[91] Although no reports have yet been published, it appears reasonable to assume that plant cytochrome P-450's may also be involved in the biosynthesis of phytosteroids.

Studies on the metabolic roles of cytochrome P-450's in molds are still few, although scattered evidence suggests that they are responsible for some key reactions in antibiotic biosynthesis. A clear example is that of m-cresol hydroxylation (a step in the biosynthesis of the antibiotic, patulin) by microsomes of *P. patulm*, which has been shown to be catalyzed by a cytochrome P-450.[40] As described above, certain step(s) in ergosterol biosynthesis in yeasts appear to be catalyzed by a cytochrome P-450. It has also been reported that the yeast *C. tropicalis*, when grown on tetradecane as carbon source, develops a cytochrome P-450-linked monooxygenase system which oxidizes the added alkane and utilizes it as a carbon and energy source for growth.[42]

Most bacterial cytochrome P-450's appear to be synthesized in the cells in response to the addition of unusual carbon sources to the growth medium, as in the case of the tetradecane-fed *C. tropicalis*. Thus, *P. putida* grown in the presence of camphor[44] and *Corynebacterium* species grown with n-octane as the sole carbon source,[45] contain high levels of cytochrome P-450. These bacterial cytochrome P-450's catalyze the first step in the oxidative degradation of the carbon source supplied. In *B. megaterium*, a cytochrome P-450 is present at low concentrations even in cells grown under normal conditions and this hemoprotein catalyzes the 15β-hydroxylation of 3-oxo-Δ^4-steroids.[47] However, it appears unlikely that the oxosteroids are converted by this reaction to metabolites which are specifically required by the organism; instead, the hydroxylation is also an initial step in their non-specific oxidative decomposition. The significance of the occurrence of several different molecular forms of cytochrome P-450 in *R. japonicum* bacteroids[46] has not yet been clarified, Since these hemoproteins are not detectable in free-living *R. japonicum* cells which are incapale of nitrogen fixation they may somehow be involved in the nitrogen fixation process. Indirect evidence has actually been presented in support of this possibility,[92] but further work is required bebore reaching a conclusion.

It is well known that a very large number of lipophilic xenobiotics including drugs, insecticides, carcinogens, food additives and environmental pollutants, are oxidatively converted to more polar metabolites by cytochrome P-450-containing monooxygenase systems in the microsomes of many animal tissues, notably those of the liver, lung, and skin. The fact that so many compounds possessing widely different structures are able to undergo oxidative transformations in these reactions originally presented an enigma to biochemists, since it was inconceivable that a single enzyme could itself exhibit such a wide substrate specificity. This enigma has, however, been partially solved on the basis of the multiplicity of cytochrome P-450 in several sources, especially in liver microsomes.[48-53] Thus, it has been shown that a cytochrome P-450 inducible in liver microsomes by phenobarbital pretreatment has a substrate specificity which is quantitatively different from that of the hemoprotein inducible with 3-methylcholanthrene.[50,53] It should be noted, however, that the enigma has not yet been completely solved in view of the very large number of xenobiotics which can be oxidized by liver microsomes.

After conversion to more polar metabolites by a cytochrome P-450-linked monooxygenase system, xenobiotics generally lose their pharmacological effects such as toxicity. On the basis of this observation, it is sometimes thought that cytochrome P-450's are a device elaborated by organisms to effect detoxication. However, the situation is by no means as simple as this, since a number of cases have been reported in which xenobiotics are converted by cytochrome P-450-linked systems to pharmacologically more active metabolites.[93-96] For instance, halogenated benzenes are converted to their arene oxides, which bind covalently to cellular macromolecules and thus cause cell damage,[95,96] and polycyclic hydrocarbons are converted to proximate and true carcinogens. It remains uncertain therefore whether oxidations of xenobiotics represent the normal function of cytochrome P-450's or whether they are unphysiological phenomena arising from the wide substrate specificity of the hemoproteins. In any case, the oxidative transformations of xenobiotics are accompanied by loss, appearance, or changes in pharmacological effects, and are therefore very important from the practical viewpoint. This is the very reason why cytochrome P-450's are being studied so actively in the field of drug metabolism.

As a conclusion to this discussion of the metabolic roles of cytochrome P-450's, it is of interest to list some of the characteristic features of cytochrome P-450-dependent metabolic sequences. First, it is apparent that all of the metabolic reactions mentioned above are due to the monooxygenase activities of the cytochrome P-450's. This suggests that the auxiliary functions of this family of hemoproteins, such as their peroxidatic and reductive activities, may not be important in connection with metabolism. Second, almost all compounds which act as substrates for cytochrome P-450 systems are more or less lipophilic. This has been explained on the basis of the fact that the active site of many cytochrome P-450's is embedded in the lipid phase of microsomal and mitochondrial membranes, and that only lipophilic substrates can penetrate to the site. However, it should be noted that the substrates of water-soluble cytochrome P-450's from bacteria are also lipophilic compounds such as camphor and paraffin hydrocarbons. Moreover, as already mentioned, the reactivity of the cytochrome P-450 in liver microsomes with added antibodies suggests that the catalytic site is exposed to the aqueous phase even in this case.[18] It seems more appropriate, therefore, to conclude that the molecular properties, especially the structure in the vicinity of the protoheme prosthetic group, of this type of hemoproteins are suitable for handling lipophilic compounds.

Another interesting feature of the cytochrome P-450-dependent metabolic pathways is that most of the sequences are not fundamental (in other words, not primitive) to the maintenance of life. Fundamental metabolic sequences such as glycolysis and protein biosynthesis should be present in all cells, but the cytochrome P-450-dependent processes are related to differentiated functions and are therefore distributed only in cells having specialized functions. In this connection, it is of interest to note that no cytochrome P-450 has been detected in fetal liver, an undifferentiated tissue,[97] or several hepatomas, dedifferentiated tissues.[98]

REFERENCES

1. J. O. Stern and J. Peisach, *J. Biol. Chem.*, **249**, 7495 (1974).
2. T. Shimizu, T. Nozawa, M. Hatano, Y. Imai and R. Sato, *Biochemistry*, **14**, 4172 (1975).
3. C. K. Chang and D. Dolphin, *J. Am. Chem. Soc.*, **97**, 5948 (1975).
4. M. Klingenberg, *Arch. Biochem. Biophys.*, **75**, 376 (1958).
5. D. Garfinkel, *ibid.*, **77**, 493 (1958).
6. T. Omura and R. Sato, *J. Biol. Chem.*, **237**, PC1376 (1962).
7. T. Omura and R. Sato, *ibid.*, **239**, 2370 (1964).
8. T. Omura and R. Sato, *ibid.*, **239**, 2379 (1964).
9. Å. Ellin, S. Jakobson, J. B. Schenkman and S. Orrenius, *Arch. Biochem. Biophys.*, **150**, 64 (1972).
10. Y. Takesue and R. Sato, *J. Biochem. (Tokyo)*, **64**, 885 (1968).
11. T. Matsubara and Y. Tochino, *ibid.*, **70**, 981 (1971).
12. R. W. Estabrook, D. Y. Cooper and O. Rosenthal, *Biochem. Z.*, **338**, 741 (1963).
13. A. P. Poland, E. Glover, J. R. Robinson and D. W. Nebert, *J. Biol. Chem.* **249**, 5599 (1974).
14. G. Betz, P. Tsai and R. Weakley, *ibid.*, **251**, 2839 (1976).
15. R. A. Meigs and K. J. Ryan, *Biochim. Biophys. Acta*, **165**, 476 (1968).
16. D. Garfinkel, *Comp. Biochem. Physiol.*, **8**, 367 (1963).
17. C. F. Strittmatter and F. T. Umberger, *Biochim. Biophys. Acta*, **180**, 18 (1969).
18. A. F. Welton, F. O. O'Neal, L. C. Chaney and S. D. Aust, *J. Biol. Chem.*, **250**, 5631 (1975).
19. B. W. Harding, S. H. Wong and D. H. Nelson, *Biochim. Biophys. Acta*, **92**, 415 (1964).

20. J. L. Purvis, J. A. Canick, J. H. Rosenbaum, J. Hologgitas and S. A. Latif, *Arch. Biochem. Biophys.*, **159**, 32 (1973).
21. T. Yohro and S. Horie, *J. Biochem. (Tokyo)*, **61**, 515 (1967).
22. H. L. Henry and A. W. Norman, *J. Biol. Chem.*, **249**, 7529 (1974).
23. J. G. Ghazarian, C. R. Jefcoate, J. C. Knutson, W. H. Orme-Johnson and H. F. DeLuca, *J. Biol. Chem.*, **249**, 3026 (1974).
24. S. Taniguchi, N. Hoshita and K. Okuda, *Eur. J. Biochem.*, **40**, 607 (1973).
25. Y. Atsuta, S. Taniguchi, K. Okuda, Y. Imai and R. Sato, *Proc. Natl. Acad. Sci. U.S.A.*, in press.
26. M. Sarte, P. V. Vignais and S. Idelman, *FEBS Lett.*, **5**, 135 (1969).
27. A. S. Khaudwala and C. P. Kasper, *Biochem. Biophys. Res. Commun.* **54**, 1241 (1973).
28. Y. Ichikawa and T. Yamano, *ibid.*, **40**, 297 (1970).
29. A Morello, W. Bleecka and M. Agosin, *Biochem. J.*, **124**, 199 (1971).
30. F. W. Plapp and J. E. Casida, *J. Econ. Entomol.*, **63**, 1091 (1970).
31. J. Capdevila, N. Ahmed and M. Agosin, *J. Biol. Chem.*, **250**, 1048 (1975).
32. J. R. M. Potts, R. Weklych and E. E. Conn, *ibid.*, **249**, 5019 (1974).
33. P. R. Rich and D. S. Bendall, *Eur. J. Biochem.*, **55**, 333 (1975).
34. O. Young and H. Beevers, *Phytochemistry*, **15**, 372 (1976).
35. P. J. Murphy and C. A. West, *Arch. Biochem. Biophys.*, **133**, 395 (1969).
36. K. M. Madyastha, T. D. Meehan and C. J. Coscia, *Biochemistry*, **15**, 1097 (1976).
37. K. M. Madyastha, J. E. Ridgway, J. G. Dwyer and C. J. Coscia, *J. Cell Biol.*, **72**, 302 (1977).
38. A. Lindenmayer and L. Smith, *Biochim. Biophys. Acta*, **93**, 445 (1964).
39. Y. Yoshida, H. Kumaoka and R. Sato, *J. Biochem. (Tokyo)*, **75**, 1201 (1974).
40. G. Murphy, G. Vogel, G. Krippahl and F. Lynen, *Eur. J. Biochem.*, **49**, 443 (1974).
41. S. H. Ambike, R. M. Baxter and N. D. Zahid, *Phytochem.*, **9**, 1959 (1970).
42. W. Duppel, J. -M. Lebeault and M. J. Coon, *Eur. J. Biochem.*, **36**, 583 (1973).
43. A. M. Vygantas and J. M. Corcoran, *Fed. Proc.*, **33**, 1233 (1974).
44. M. Katagiri, B. N. Ganguli and I. C. Gunsalus, *J. Biol. Chem.*, **243**, 3543 (1968).
45. G. Cardini and P. Jurtchuk, *ibid.*, **245**, 2789 (1970).
46. C. A. Appleby, *Biochim. Biophys. Acta*, **147**, 399 (1967).
47. A. Berg, J. -Å. Gustafsson, M. Ingelman-Sundberg and K. Carlström, *J. Biol. Chem.*, **251**, 2831 (1976).
48. Y. Imai and R. Sato, *Biochem. Biophys. Res. Commun.*, **60**, 8 (1974).
49. C. Hashimoto and Y. Imai, *ibid.*, **68**, 821 (1976).
50. D. Ryan, A. Y. H. Lu, S. West and W. Levin, *J. Biol. Chem.*, **250**, 2157 (1975).
51. J. C. Kawalek, W. Levin, D. Ryan, P. E, Thomas and A. Y. H. Lu, *Mol. Pharmacol.*, **11**, 874 (1975).
52. A. P. Alvares, D. R. Bickers and A. Kappas, *Proc. Natl. Acad. Sci. U.S.A.*, **70**, 1321 (1973).
53. A. F. Welton and S. D. Aust, *Biochem. Biophys. Res. Commun.*, **56**, 898 (1974).
54. S. Takemori, K. Suhara, S. Hashimoto, H. Sato, T. Gomi and M. Katagiri, *ibid.*, **63**, 588 (1975).
55. S. Takemori, H. Sato, T. Gomi, K. Suhara and M. Katagiri, *ibid.*, **67**, 1151 (1975).
56. R. Goewert, J. Mitchell, C. A. Appleby and K. Dus, *Fed. Proc.*, **34**, 623 (1975).
57. T. Omura, E. Sanders, R. W. Estabrook, D. Y. Cooper and O. Rosenthal, *Arch. Biochem. Biophys.*, **117**, 660 (1966).
58. M. Bryson and M. L. Sweat, *J. Biol. Chem.*, **243**, 2799 (1968).
59. L. Sagan, *J. Theoret. Biol.*, **14**, 225 (1967).
60. A. Y. H. Lu and M. J. Coon, *J. Biol. Chem.*, **243**, 1331 (1968).
61. Y. Imai and R. Sato, *Biochem. Biophys. Res. Commun.*, **60**, 8 (1974).
62. T. Iyanagi and H. S. Mason, *Biochemistry*, **12**, 3789 (1973).
63. H. W. Strobel, A. Y. H. Lu, J. Heidema and M. J. Coon, *J. Biol. Chem.*, **245**, 4851 (1970).
64. K. Ichihara, E. Kusunose and M. Kusunose, *Biochim. Biophys. Acta*, **239**, 178 (1971).

65. Y. Yoshida and H. Kumaoka, *J. Biochem. (Tokyo)*, **78**, 785 (1975).
66. J. R. Gillette, D. C. Davis and H. A. Sasame, *Ann. Rev. Pharmacol.*, **12**, 57 (1972).
67. E. G. Hrycay and R. A. Prough, *Arch. Biochem. Biophys.*, **165**, 331 (1974).
68. B. S. Cohen and R. W. Estabrook, *ibid.*, **143**, 37 (1971).
69. A. Hildebrandt and R. W. Estabrook, *ibid.*, **143**, 66 (1971).
70. H. A. Correia and G. J. Mannering, *Mol. Pharmacol.*, **9**, 455 (1973).
71. G. J. Mannering, H. Kuwahara and T. Omura, *Biochem. Biophys. Res. Commun.*, **62**, 150 (1975).
72. Y. Imai and R. Sato, *ibid.*, **75**, 420 (1977).
73. E. G. Hrycay and P. J. O'Brien, *Arch. Biochem. Biophys.*, **157**, 7 (1973).
74. E. G. Hrycay, H. G. Jonen, A. Y. H. Lu and W. Levin, *ibid.*, **166**, 145 (1975).
75. A. D. Rahimtula and P. J. O'Brien, *Biochem. Biophys. Res. Commun.*, **60**, 440 (1974).
76. J. -Å. Gustafsson, E. G. Hrycay and L. Ernster, *Arch. Biochem. Biophys.*, **174**, 440 (1976).
77. J. R. Gillette, J. J. Kamm and H. A. Sasame, *Mol. Pharmacol.*, **4**, 541 (1968).
78. P. H. Hernandez, P. Mazel and J. R. Gillette, *Biochem. Pharmacol.*, **16**, 1877 (1967).
79. M. Sugiura, K. Iwasaki and R. Kato, *Mol. Pharmacol.*, **12**, 746 (1976).
80. D. Y. Cooper, S. Levin, S. Narasimhulu, O. Rosenthal and R. W. Estabrook, *Science*, **147**, 400 (1965).
81. G. F. Gibbons and K. A. Mitropouls, *Eur. J. Biochem.*, **40**, 269 (1973).
82. K. T. W. Alexander, K. A. Mitropouls and G. F. Gibbons, *Biochem. Biophys. Res. Commun.*, **60**, 460 (1974).
83. G. S. Boyd, A. M. Grimwade and M. E. Lawson, *Eur. J. Biochem.*, **37**, 334 (1973).
84. E. R. Simpson and G. S. Boyd, *Biochem. Biophys. Res. Commun.*, **24**, 110 (1966).
85. R. A. Meigs and K. J. Ryan. *J. Biol. Chem.*, **246**, 83 (1971).
86. E. A. Thompson, Jr. and P. K. Siiteri, *ibid.*, **249**, 5373 (1974).
87. M. H. Bhattacharyya and H. F. DeLuca, *Arch. Biochem. Biophys.*, **160**, 58 (1974).
88. F. Wada, H. Shibata, M. Goto and Y. Sakamoto, *Biochim. Biophys. Acta*, **162**, 518 (1968).
89. Å. Ellin, S. Orrenius, Å. Pillotti and C. -G. Swahn, *Arch. Biochem. Biophys.*, **158**, 597 (1973).
90. P. J. Murphy and C. A. West, *ibid.*, **133**, 395 (1969).
91. S. H. Ambike, R. M. Baxter and N. D. Zabid, *Phytochemistry*, **9**, 1953 (1970).
92. W. L. Kretovich, S. S. Melik-Sarkissian, M. V. Raikchinstein and A. I. Archakov, *FEBS Lett.*, **44**, 305 (1974).
93. D. M. Jerina and J. W. Daly, *Science*, **185**, 573 (1974).
94. J. R. Gillette, J. R. Mitchell and B. B. Brodie, *Ann. Rev. Pharmacol.*, **14**, 271 (1974).
95. J. W. Daly, D. M. Jerina and B. W. Witkop, *Experientia*, **28**, 1129 (1972).
96. T. Shimada and R. Sato, *Biochem. Pharmacol.*, in press.
97. G. Dallner, P. Siekevitz and G. E. Palade, *Biochem. Biophys. Res. Commun.*, **20**, 135 (1965).
98. K. Ikeda, M. Hozumi and T. Sugimura, *J Biochem. (Tokyo)*, **58**, 595 (1965).

CHAPTER 3

Molecular Properties

3.1. Purification and Chemical Properties[*1]
    3.1.1. Purification of Hepatic Microsomal P-450
    3.1.2. Chemical Properties
    3.1.3. Conversion of P-450 to P-420
    3.1.4. Interaction of P-450 with Isocyanides
3.2. Optical Properties[*2]
    3.2.1. Development of Research
    3.2.2. Difference Spectra
    3.2.3. Absolute Absorption Spectra
    3.2.4. Extinction Coefficients
    3.2.5. Photochemical Action Spectrum of P-450
3.3. Magnetic Properties[*3,4]
    3.3.1. History and an Outline of Microsomal Fe_x
    3.3.2. Short Survey of Ligand Field Theory
    3.3.3. High Spin P-450 Induced by 3-Methylcholanthrene
    3.3.4. EPR Signal Changes of P-450 with Various Ligands or Substrates
    3.3.5. EPR Study of P-450 Conversion to P-420
    3.3.6. NMR Study of P-450
    3.3.7. Magnetic and Molecular Spectroscopic Investigations on P-450 Other than EPR and NMR Studies

3.1 PURIFICATION AND CHEMICAL PROPERTIES

3.1.1 Purification of Hepatic Microsomal P-450

Cytochrome P-450 was first found[1,2] and recognized as a hemoprotein[3,4] in liver microsomes. Early studies on its optical and magnetic

---

[*1] Yoshio IMAI, Institute for Protein Research, Osaka University, Suita, Osaka 565, Japan
[*2] Shigeo HORIE, School of Medicine, Kitasato University, Sagamihara, Kanagawa 228, Japan
[*3] Toshio YAMANO, 1st Department of Biochemistry, Osaka University Medical School, Kita-ku, Osaka 530, Japan
[*4] Tetsutaro IIZUKA, Department of Biochemistry, Keio University School of Medicine, Shinjuku-ku, Tokyo 160, Japan

properties were therefore carried out mainly using liver microsomes, and successfully disclosed its unique characteristics. After the finding of P-450 several attempts were made to purify hepatic microsomal P-450 to a homogeneous state proving this cytochrome to be an unusually difficult protein to purify. However, success has recently been reported in obtaining homogeneous preparations.[5-11]

In this section, progress in the purification of hepatic microsomal P-450* will be described. Cytochromes from other sources will be dealt with elsewhere.

A. Difficulty in solubilization and purification

The main reasons for the difficulty in purification lay in the high instability of the cytochrome on solubilization and in the difficulty of separation of integral proteins of microsomal membranes from one another (these are likely to associate by hydrophobic bonding) without denaturation. Unsaturated fatty acids involved in the phospholipid as a constituent of microsomal membranes are oxidizable with air to produce lipid peroxides, accompanied by destruction of the heme.[4,14] Thus, P-450 disappears gradually without corresponding formation of P-420, when microsomes are digested aerobically with proteases. On anaerobic treatment, the cytochrome is recovered without solubilization as its denatured form, P-420, which is rather less stable than P-450, easily decomposing and disappearing under aerobic conditions.[4,14] Since P-450 is firmly bound in microsomal membranes and is one of the least solubilizable proteins in microsomes against the general procedures used for solubilizing membrane proteins, P-450 could be solubilized only by treatments which caused disruption of the hydrophobic interactions and accompanying denaturation, i.e. conversion to P-420.[4,14] These are the reasons why P-450 proved too labile to be purified in native form for a long time.

B. Purification of P-420

Since the treatments to solubilize P-450 always induced its conversion to P-420, until an effective protecting agent was found, attempts

* "P-450" was first used to designate hepatic microsomal CO-binding pigments obtained from untreated animals.[3,4] The pigment from methylcholanthrene-(MC)-treated animals has a number of different properties from that from phenobarbital(PB)-treated or untreated animals, and has therefore often been called P-448[12] or P_1-450.[13] Moreover, evidence has recently been accumulated for the existence of multiple molecular species of P-450 in liver microsomes. Thus, the name "cytochrome P-450" will be used inclusively in this section for all pigments, the CO-compound of which has an absorption peak at *around 450 nm*, as a group name, independent of origin.

were first made to purify the cytochrome in its "solubilized form", P-420. Thus, P-420 was purified by Omura and Sato[14] to a content of 6–7 nmole protoheme per mg of protein from rabbit liver microsomes. Their method involved fractionation with ammonium sulfate, treatment with calcium phosphate gel and gel filtration through Sephadex G-100 after digestion of the microsomes with Steapsin to remove cytochrome b_5 followed by solubilization with heated snake venom (phospholipase A) and deoxycholate under anaerobic conditions. Although P-420 is a denatured form of P-450, studies on some of its basic properties made the hemoprotein nature of this cytochrome clear.

C. Protecting agents

Ichikawa and Yamano[15] found that glycerol and other polyols are able to stabilize P-450, preventing its conversion to P-420. This gave a clue for the later remarkable progress in the purification of P-450, which has rendered the solubilization and purification of hepatic microsomal P-450 in native form possible. Glycerol is now being used widely as a protecting agent for P-450 during its solubilization, purification and storage.

EDTA, dithiothreitol (DTT)[16] and butylated hydroxytoluene[17] were found to be effective in protecting P-450 against decomposition by lipid peroxidation with air, and are therefore also added to the medium on solubilization and on purification, especially in the early stages.

D. P-450 particles

Considering the difficulty in obtaining a stable preparation of soluble P-450 which has the spectra of native P-450, attempts were made to prepare particles containing P-450 as the sole heme constituent which would be suitable for measurement of the absolute spectra of P-450. When microsomes are digested anaerobically with proteases, cytochrome b_5, which is a heme constituent of the microsomes and present in excess over the P-450, can be almost completely solubilized, leaving the CO-binding pigment still attached to the membrane residues, although P-450 is converted to P-420 to a considerable extent[4,14]. Nagarse (protease from *Bacillus subtilis*) was found to be the most effective protease examined for removing cytochrome b_5 with the least conversion of P-450 to P-420.[18] On proteolytic digestion, glycerol was found to act as a protecting agent against such conversion.[18] Thus, Nishibayashi and Sato[18] successfully obtained P-450 particles from liver microsomes of PB-treated rabbits by anaerobic digestion with Nagarse in the presence of glycerol. The particles contained about 5.5 nmole P-450 per mg of protein, which was almost identical with the content in the original microsomes. The content of P-420 and of cytochrome b_5 in the particles represented as little as 1% and 5% of the total heme, respectively.

Ichikawa and Yamano[19] also described a method for isolating a similar preparation by tryptic digestion of rabbit liver microsomes.

Although soluble preparations of P-450 were obtained from P-450 particles by treatment with detergents,[20-22] it is not known whether a portion of the protein moiety is cleaved on proteolysis from the native protein or not.

E. Solubilization by detergents and purification

Lu and Levin[23] have given a detailed review of methods developed for the solubilization, purification of hepatic microsomal P-450, with the focus on the resolution and reconstruction of the hydroxylase system. In this section, recent progress in the purification of P-450 will be described briefly.

Miyake et al.[24] first succeeded in the solubilization of P-450 with a detergent, followed by partial purification without essential modification of the optical and magnetic properties of P-450, although they referred to their preparation as *submicrosomal particles*. The P-450 was solubilized from liver microsomes of PB-treated rabbits with the non-ionic detergent, Lubrol WX, in the presence of glycerol. The solubilization step was followed by ammonium sulfate fractionation and column chromatography on DEAE-Sephadex. The preparation contained 2.5 nmole P-450 per mg of protein, while the contents of both P-420 and cytochrome b_5 were less than 1% of this figure. The total yield of P-450 from the original microsomes was only 5%. NADH-ferricyanide reductase did not separate from the P-450 and considerable amounts of phospholipid still remained unremoved.

Fujita et al.[25] used the non-ionic detergent, Triton N-101, for the solubilization and purification of P-450 from liver microsomes of PB- and MC-treated rats. Their solubilization was carried out in the presence of glycerol and was followed by ammonium sulfate fractionation, column chromatography on DEAE-cellulose, and treatment with alumina gel-C_7. The content of cytochromes for both the PB- and MC-treated animals was calculated to be about 5 nmole per mg of protein. These values, however, may be questionable since extinction coefficients of 58 and 78 $cm^{-1} \cdot mM^{-1}$ were used for the cytochromes of PB- and MC-treated animals, respectively, instead of the generally accepted value of 91 $cm^{-1} \cdot mM^{-1}$. The preparation still contained NADH-ferricyanide reductase and the total recovery of cytochromes from the original microsomes was about 15%.

Sato et al.[26] used Emulgen 911 for the solubilization and partial purification of P-450 from liver microsomes of PB-treated rabbits. Their partially purified preparation of P-450 obtained by column chromatography on DEAE-cellulose showed a specific content of over 10 nmole per mg of protein, but was not free of cytochrome b_5 or P-420.

Lu and Coon[16] first reported the solubilization, resolution and reconstruction of a microsomal hydroxylase system, which consisted of cytochrome P-450, NADPH-cytochrome c reductase and lipid factor, by using an ionic detergent. Rabbit liver microsomes were treated with deoxycholate in the presence of glycerol and DTT in citrate buffer, and the resulting soluble fraction was resolved into three components by column chromatography on DEAE-cellulose. One of these components was identified as P-450, but considerable and variable amounts of P-450 were converted to P-420 by these procedures since deoxycholate is a rather powerful reagent for inducing the conversion; 20 to 60% of the cytochrome in the eluted fraction was present in the form of P-420.[16,23]

Lu and Levin[23] described the elution profile and separation of components of the microsomal electron transport system on DEAE-cellulose with detergents, comparing the ionic detergent, deoxycholate, with the nonionic detergents, Triton N-101 and Lubrol WX. With deoxycholate, the cytochrome was well separated from the phospholipid but incompletely from the reductase. In contrast, with the non-ionic detergents, the cytochrome was not separated from the phospholipid but was generally well separated from the reductase.

Lu et al.[27] used cholate for the solubilization of P-450 from liver microsomes of PB- and MC-treated rats. The solubilization procedure was carried out in the presence of glycerol and DTT, and was followed by repeated fractionation with ammonium sulfate and treatment with calcium phosphate gel. Lu and Levin, thus, obtained a partially purified preparation of the cytochromes having a specific content of 5.0 to 5.5 nmole per mg of protein.[28–30] The same procedure yielded P-450 at 2.4 nmole per mg of protein in uninduced rats.[29,30] About 90% of the total phospholipid was removed but cytochrome b_5 was still present. The total recovery of P-450 from the original microsomes was 4 to 8%. Levin et al.[30] used the nonionic detergent, Emulgen 911, for further purification of this partially purified preparation by column chromatography on DEAE-cellulose and then on Sephadex LH 20 to remove excess Emulgen 911. The final preparation contained 9 to 11 nmole of the cytochromes per mg of protein and was free of cytochrome b_5, NADPH-cytochrome c reductase and phospholipid, although about 10% of the cytochrome was in the form of P-420 judging from its CO-difference spectrum.

Van der Hoeven et al.[31] reported a method for the partial purification of P-450 of higher specific content from liver microsomes of PB-treated rabbits, obtaining a better recovery rate. The microsomes were extracted with pyrophosphate and then treated with cholate in the presence of glycerol, EDTA, DDT and butylated hydroxytoluene. The solubilized P-450 was fractionated by precipitation with polyethylene glycol 6000, submitted to column chromatography on DEAE-cellulose in the presence of the non-ionic detergent, Renex 690, and then treated with Amberlite XD-2 and calium phosphate gel. The final preparation had a specific content of about 13 nmole P-450 per mg of protein, with an overall yield of 22% from the orginal microsomes. It was free of cytochrome b_5, and contained no significant amounts of P-420, NADPH-cytochrome c reductase or NADH-ferricyanide reductase.

Imai and Sato[32] described a simple method for the partial purification of P-450 from liver microsomes of PB-treated rabbits. The P-450 was solubilized by treatment with cholate in the presence of glycerol, DTT and EDTA, adsorbed on a column of ω-amino-n-octyl derivatives of Sepharose 4B, and then eluted with a medium containing Emulgen 913. With some modification of the solubilization conditions[33] an overall yield of about 50% was obtained, i.e. when fractions with a specific content of P-450 of over 9 nmole per mg of protein were pooled. Neither P-420, cytochrome b_5 nor NADPH-cytochrome c reductase was present in this latter preparation.

In 1974, two groups independently reported the successful purification of hepatic microsonal P-450 of PB-treated rabbits to a gel-electrophoretically homogeneous state from partially purified preparations.[5,8] Imai and Sato[5] obtained a preparation of P-450 having a specific content of about 17 nmole per mg of protein, at an overall yield of about 10%, by column chromatography on hydroxylapatite and then on CM-Sephadex C-50. (Details of a slight modified method will also be described later.) van der Hoeven et al.[8] obtained a preparation of P-450 having a specific content of about 17 nmole per mg of protein, at an overall yield of about 3%. In this case, the preparation fractionated on DEAE-cellulose and treated with calcium phosphate gel was submitted to further purification by column chromatography on hydroxylapatite and treatment with calcium phosphate gel, followed by a repeat of this same procedure.

Such purified preparations of P-450 exhibit only a single polypeptide band on SDS-urea-polyacrylamide gel electrophoresis, and an apparent molecular weight of about 50,000 was estimated for the cytochrome from a comparison of its mobility with that of marker proteins.[5,6,8,9]

This value is considerably lower than the minimal molecular weight calculated from the specific content of P-450 in the purified preparations. Several possible explanations of this discrepancy have been discussed,[5] and the presence of a considerable amount of apocytochrome in the preparations appears to be the most likely case.

By modifications of their methods for the purification of P-450 from PB-treated animals, both groups further obtained apparently homogeneous preparations of another molecular species of the cytochrome from MC-, β-naphthoflavone- or PB-treated rabbits.[6,7,9] When the procedures for purification from liver microsomes of rabbits are applied to rats, the situation appears to be somewhat different; the PB- and MC-induced P-450's behave in a similar way.

Lu's groups also obtained highly purified preparations of the cytochrome from PB- or MC-treated rats[10] and MC-treated rabbits[11] by fractionation of the partially purified preparations[30] on a CM-cellulose column.

F. Purification of P-450 to a homogeneous state

In the method developed by Imai and Sato,[5] the practical procedures are much simpler and the overall yield much better than in the method of Coon's group.[8,9] Furthermore, the preparation obtained in the former case contained practically no detergent. This method will be described next in some detail.

Liver microsomes prepared from PB-treated rabbits in isotonic KCl containing 10 mM EDTA were suspended at a protein concentration of about 1.5 mg/ml in 100 mM potassium phosphate buffer (pH 7.25) containing 0.6% (w/v) sodium cholate, 20% (v/v) glycerol, 1 mM DTT and 1 mM EDTA (all final concentrations). (Potassium phosphate buffer solutions, pH 7.25, containing 20% glycerol were employed throughout; they will be referred to simply hereafter as 100 mM buffer, etc.) The suspension was allowed to stand for 30 min and then centrifuged at 77,000 g for 120 min. The supernatant fraction was applied to a column of ω-amino-n-octyl Sepharose 4B (2.7 × 32 cm, for 1.5 g of original microsomal protein) which had been equilibrated with the same buffer. After washing the column with 3 times the column volume of 100 mM buffer containing 0.5% sodium cholate, 1 mM DTT and 1 mM EDTA, the tightly absorbed P-450 was eluted with 100 mM buffer containing 0.4% sodium cholate, 1 mM DTT and 1 mM (0.08 %) Emulgen 913, and fractions with high specific contents of P-450 were pooled. On elution, it was necessary to keep the flow rate at 20 ml/h or less, since if the flow rate was too fast, both the specific content and recovery of P-450 in the peak fractions were reduced. The pooled fractions were diluted 4-fold with 20% glycerol and adsorbed onto a column (3.2 × 10 cm) of hydroxylapatite equilibrated with 25 mM buffer. After washing the column with 40 mM buffer containing 0.2% Emulgen 913, elution was conducted step-wise with 100 mM buffer and then with 150 mM buffer, both containing 0.2% Emulgen 913. The peak fractions of the 100 mM buffer eluate were combined and diluted 5-fold with 20 % glycerol containing 0.2% Emulgen 913. The diluted solution was then applied to a CM-Sephadex column (2.1 × 15 cm) equilibrated with 20 mM buffer containing 0.2% Emulgen 913. This column was washed with a small amount of the same buffer, and then with 50 mM buffer containing the Emulgen. A slight reddish color leaked out from the column. P-450 was then eluted sharply as a highly concentrated solution by increasing the buffer concentration to 100 mM. At this step, it was necessary to keep the flow rate as slow as possible in order to obtain good results. The peak fractions of the 100 mM buffer eluate showed a specific content exceeding 17 nmole per mg of protein and exhibited a single polypeptide band when subjected to polyacrylamide gel electrophoresis in the presence of sodium dodecyl sulfate, urea and β-mercaptoethanol. Practically all of the Emulgen in the sample could be removed by a second CM-Sephadex column; the fraction was diluted 5-fold with 20% glycerol and the solution was applied to a CM-Sephadex column equilibrated with 20 mM buffer. The column was washed with 20 mM buffer until no absorption at 276 nm due to Emulgen was detectable in the eluate. P-450 was then eluted with 150 mM buffer. It was necessary to carry out this step on a rather small scale in order to remove the Emulgen efficiently and to avoid apo-cytochrome formation.

Table 3.1 gives a summary of the results of a typical purification experiment, and the preparation so obtained can be stored at −70°C

TABLE 3.1. Purification of P-450 from liver microsomes of PB-treated rabbits[†]

Fraction	Protein (mg)	Cytochrome P-450		
		Total content (nmole)	Specific content (nmole/mg protein)	Recovery (%)
Microsomes	1490	3860	2.57	100
Solubilized supernatant	1340	3790	2.82	98
Aminooctyl column eluate	213	2150	10.1	56
Hydroxylapatite eluate	67.7	984	14.5	25
1st CM-Sephadex eluate	37.8	662	17.5	17

[†] The step of 2nd CM-Sephadex was carried out part by part. For example, when 300 nmole of the 1st CM-Sephadex eluate was applied to the 2nd CM-Sephadex column (1.2 × 12 cm), 230 nmole P-450 (specific content, 17.5) was recovered in the peak fractions. The whole recovery of this step was 92%, and the specific content remained unchanged.

for at least several months, even without removing air, provided that glycerol is present. However, repeated freezing and thawing (especially in the absence of the Emulgen) results in the formation of aggregates. The Emulgen appears to act as a protecting agent for the P-450 in the purified preparation. On incubation of the CO-compound of the purified preparation, gradual conversion of P-450 to P-420 occurs even with no addition of reagents which induce the conversion. The presence of Emulgen 913 serves to slow down the rate of conversion.[34]

It is now generally accepted that there exist multiple species of P-450 which differ from one another in many respects. A P-450 preparation which has clearly different catalytic properties, can be obtained from the 150 mM buffer eluate of a hydroxylapatite column,[34] although it is not yet homogeneous. Major parts of the P-450 present in liver microsomes from MC-treated rabbits (Table 3.2) were still retained

TABLE 3.2. Purification of P-450 from liver microsomes of MC-treated rabbits†

Fraction	Protein (mg)	Cytochrome P-450		
		Total content (nmole)	Specific content (nmole/mg protein)	Recovery (%)
Microsomes	1920	7210	3.75	100
Solubilized supernatant	1510	6690	4.42	93
Aminooctyl column eluate	167	2130	12.8	30
Hydroxylapatite eluate	57.5	1010	17.6	14
1st CM-Sephadex eluate	35.0	620	17.7	9

† The step of 2nd CM-Sephadex was cairred out part by part in the same way as described in Table 3.1.

on the column of ω-amino-n-octyl Sepharose after washing with 1 mM (0.08%) Emulgen 913, and were eluted out by increasing the Emulgen concentration to 2.5 mM (0.2%).[6] This species of the cytochrome, which has an absorption maximum at 448 nm in the CO-compound, could not be washed out from a hydroxylapatite column with 150 mM buffer containing 0.2% Emulgen 913. Addition of 0.1% cholate to the buffer was required for elution of the cytochrome from the column.[6] At the step of CM-Sephadex, the cytochrome was eluted out at a higher concentration of phosphate.[6] The preparation thus obtained exhibits an absorption spectrum of the high spin state in the oxidized form. Starting from liver microsomes of PB-treated rabbits the same procedure was also applicable to obtain a gel-electrophoretically homogeneous preparation of a P-450 species identical with (or closely similar to) the MC-induced one, although the amounts obtainable were far less.[7]

Additional points regarding the method described above are as follows.

(1) *Use of ω-amino-n-octyl Sepharose for the first step of the purification.* In early reports on the purification of P-450, the procedures for separation of the cytochrome from other proteins were unavoidably accompanied by decomposition of the cytochrome due to its high instability. The specific content thus did not increase appreciably and the recovery of P-450 was poor. In contrast, almost 90% of the P-450 applied to an ω-amino-n-octyl Sepharose column is recovered as a whole and more than half the total P-450 recovered is concentrated into peak fractions, which have a specific content of about 10 nmole per mg of protein. That is to say, a P-450 preparation which is almost 50% pure can be obtained in the first step of purification at a yield of about 50%. Moreover, the P-450 obtained in the sample after using this column is very stable, provided that glycerol is present.

(2) *Selection of the kind and concentration of detergents at each step of purification.* As Lu and Levin have pointed out,[23] cholate is a good choice for solubilization of P-450 and for its purification, at least at the first step. Both the concentration of cholate and the ratio of cholate to protein are important for obtaining a good recovery of P-450 from the ω-amino-n-octyl Sepharose column. When either value is too low, certain amounts of the P-450 pass through the column without being adsorbed, probably due to incomplete solubilization.[34] The phosphate concentration also affects the efficiency of the solubilization. Moreover, the concentrations of cholate and Emulgen 913 are important for the effective separation of P-450 from the other microsomal proteins on this column. For example, at a concentration of 0.6% cholate, both P-450 and cytochrome b_5 migrate very slowly together even in the absence of the Emulgen. A too great an increase in the concentration of the Emulgen, even at a concentration of 0.4% cholate, results in the leakage of cytochrome b_5 with the elution of P-450. Use of two appropriate concentrations of the Emulgen can afford separate elution of at least two molecular species of P-450 without the leakage of cytochrome b_5. Changing the detergent from cholate to the Emulgen is necessary for the purification of P-450 from PB-treated animals on a hydroxylapatite column with an increased concentration of phosphate. On the other hand, P-450 from MC-treated animals can be eluted with cholate after removal of the other proteins by washing with an increased concentration of phosphate in the presence of the Emulgen as a detergent.

3.1.2 Chemical Properties

A. Prosthetic group

Since protoheme is the only heme present in liver microsomes[4] or in partially purified P-420[14] on the basis of the alkaline pyridine-hemochromogen spectrum, and the P-450 content is equal to that of the microsomal protoheme which is present in excess of the amount associated with cytochrome b_5, it has generally been accepted that the heme moiety of P-450 is ferriprotoporphyrin IX. However, the spectral anomaly of P-450 continued to occasion some doubt as to the validity of the above assumption until the nature of the heme moiety was fully characterized. Maines and Anders[35] successfully confirmed its validity using a reductive degradation technique for heme identification. The heme moiety of P-450 was removed by acid-acetone treatment of CO-binding particles, devoid of cytochrome b_5, which were prepared by steapsin digestion of liver microsomes from MC-treated and untreated rats. When the products of the reductive degradation and their derivatives were examined by gas chromatography, mass spectrometry and a combination of both, no differences were observed in the degradation products of the heme from either of the microsomes or the hemoglobin of corresponding animals which served as a control. This indicated that the heme of P-450 from both MC-treated and untreated animals was identical, and that it is the same as that of hemoglobin, viz. ferriprotoporphyrin IX.

Gunsalus et al.[36,37] determined the prosthetic group of bacterial cytochrome P-450 to be ferriprotoporphyrin IX. Heme was removed from purified P-450 by acid-acetone treatment, and was identified from its pyridine-hemochromogen spectrum and by co-chromatography of authentic and isolated samples.[37] Moreover, the spectra and catalytic activity of P-450 could be satisfactorily restored on addition of equimolar protoheme to the apo-cytochrome followed by cysteine glycerol treatment.[37]

B. Molecular weight

Purified P-450 from liver microsomes yields only a single polypeptide band on SDS-urea-polyacrylamide gel electrophoresis.[5-10] The molecular weight of the polypeptide chain of P-450 as estimated by the gel electrophoresis method, differs somewhat depending on the species and on the pretreatment of the animals, as shown in Table 3.3. The values reported for the major P-450's from PB- and MC-treated animals are about 50,000 and 55,000, respectively.[6,9-11,34,38,39] A molecular

TABLE 3.3. Molecular weights of hepatic microsomal P-450

Source	S. C. of preparation (nmole/mg protein)	Molecular weight	Ref.
Rabbit, PB-treated	17–18	50,000	6
Rabbit, PB-treated	14–21	49,000	9
Rat, PB-treated	15–16	49,000	38
Rat, PB-treated	12–13	48,000	10
Rabbit, PB-treated	~18	54,000	7
Rabbit, MC-treated	17–19	54,000	6
Rabbit, PB-treated NF-treated†	13–17	55,000	9
Rabbit, MC-treated	15–17	51,000	11
Rat, MC-treated	14–16	53,000	10
Mouse, PB-treated	12	50,000	39
	15	56,000	39

† NF = β-naphthoflavone.

weight range of 44,000 to 46,000 has been estimated for bacterial P-450 by several methods.[36,40,41] Two molecular species of cytochrome P-450, viz. P-450$_{11\beta}$ and P-450$_{scc}$, have been purified from bovine adrenal mitochondria.[42,43] Their molecular weights have been estimated at 43,000[42] and 46,000,[43] respectively, by sedimentation equilibrium of guanidine-treated proteins or at 46,000 and >46,000, respectively, by SDS-urea-gel electrophoresis.[42] Although the monomeric molecular weight of P-450 from hepatic microsomes or adrenal mitochondria is about 50,000, the molecular weight of the functional unit in the membranes, as oxygenase, is not clear. The apparent molecular weight of purified P-450 from liver microsomes in an aqueous medium has been determined to lie in the range, 250,000 to 600,000.[17,31,44,45] Also it is not unreasonable to suppose that the value may be largely variable dependent on the conditions, which can affect the polymeric states of amphipatic proteins, since hepatic P-450 belongs to the typical integral protein of microsomal membranes. The apparent molecular weight of adrenal mitochondrial P-450 in an aqueous medium has been reported as 850,000[46] (when containing 8 hemes and 16 identical polypeptide chains), and it is dissociated to subunits of 53,000 daltons in 6 M guanidine hydrochloride.[47]

C. Amino acid composition

The amino acid compositions of two molecular species of hepatic microsomal P-450 are given in Table 3.4. Haugen and Coon[9] have also reported similar results for their preparation. The cytochrome contains relatively high contents of leucine and phenylalanine. By the procedure of Capaldi and Vanderkooi,[45] 55–57% of the amino acid

TABLE 3.4. Amino acid compositions of P-450

	Rabbit liver microsomes		Pseudomonas putida[48]	Rhizobium japonicum[49]
	PB-treated	MC-treated		
Asp	} 35	} 38	27	} 33
Asn			9	
Thr	23	23	19	14
Ser	27	40	21	20
Glu	} 39	} 46	42	} 54
Gln			13	
Pro	30	37	27	25
Gly	31	29	26	33
Ala	21	31	34	54
Val	26	31	24	25
Met	8	7	9	8
Ile	22	21	24	14
Leu	52	51	40	46
Tyr	12	8	9	6
Phe	36	31	17	17
His	19	19	12	12
Lys	12	11	13	14
Arg	32	40	24	29
Try	2	9	1	1
Cys/2	3	5	6	2
Total	430	477	397	407

(Data from ref. 48, 49 reproduced by kind permission of Academic Press, Inc., U. S. A.)

residues are nonpolar, providing an explanation for the tendency of the protein to aggregate in detergent-free aqueous media. This value is larger than that for cytochrome b_5 (44–48%) and NADPH-cytochrome c reductase (41%), about same as that for NADH-cytochrome b_5 reductase (58%) and epoxide hydrase (58%), but is less than that for fatty acid desaturase (62%). In other words, the content of hydrophobic amino acid residues in hepatic microsomal P-450 is, as a whole, rather usual for a microsomal membrane protein, in spite of the fact that a lipophilic nature is characteristic of the substrate for monooxygenase, the cytochrome. Dus et al. have reported the amino acid compositions of bacterial P-450 from camphor-induced cells of Pseudomonas putida[41] and of hepatic microsomal P-450 from PB-treated rabbits (preparation 90 % pure), and compared them.[48] Amino acid analysis of Rhizobium japonicum P-450 has also been carried out.[49] A high similarity can be seen between the four cytochromes from the different sources in their amino acid compositions (see Table 3.4), in spite of the unrelated substrate specificities of the cytochromes.

Both bacterial P-450 and hepatic microsomal P-450 contain carbohydrate.[9,41,49] Glucosamine was first recognized as a component of bacterial P-450 (0.7 residues per 45,000 g of protein).[41] The presence of an additional carbohydrate component in the cytochrome has been found but its nature, locus and mode of linkage to the polypeptide chain are not yet known.[41] It has been reported[9] that microsomal P-450 contains 1 mole of glucosamine and 2 mole of mannose per mole of the cytochrome.

The isoelectric point for bacterial P-450 as determined by electro-focusing experiments is at pH 4.55, which is in agreement with the excess of acidic over basic residues.[41] An increase of 0.12 pH units in the isoelectric point was observed on complex formation with camphor.[41] The pH for microsomal P-450 is more acidic (pH 4.2) than that for bacterial P-450 in spite of the former's fewer acidic than basic residues, a fact which probably results from acidic carbohydrate units.[48]

Dus et al.[41] examined the amino terminus and carboxyl terminus of bacterial P-450. Asparagine or aspartic acid was determined as the amino terminus by the DNP method, although its recovery is known to be poor (15%). A DNP-octapeptide of composition DNP–Asx(–ε-DNP–Lys, Ser, Gly, Val, Ile, Leu) was obtained by mild acid hydrolysis of DNP derivatives of P-450 followed by extraction and purification. Valine was determined as the carboxyl terminus by hydrazinolysis in almost quantitative recovery. Digestion of bacterial P-450 with carboxypeptidase A gave the carboxyl terminus sequence as (Asp, Leu, Met, Ile, Glu, Lys)–Ala–Leu–Thr–Val–OH. Haugen and Coon[9] determined the carboxyl terminus of microsonal P-450 from PB-treated and β-naphthoflavone-treated rabbits as arginine and lysine, respectively. The amino terminal sequence of P-450 from PB-treated rabbits was determined; the terminus, methionine and the region, hydrophobic.[50]

Partial degradation of P-450 by selective chemical cleavage with BrCN under suitable conditions yields a small hemepeptide which can be purified without loss of heme.[48] The amino acid compositions of the smallest hemepeptides thus obtained from bacterial P-450 and microsomal P-450 are shown in Table 3.5.[48] A single residue of histidine and cysteine occurs in each hemepeptide, and these could be candidates for the axial ligands for iron in P-450 heme. Although the hemepeptide of microsomal P-450 is smaller by 10 residues, a general similarity in composition is seen, suggesting a structural similarity in the heme region between the two cytochromes.

D. Thiolate anion as a possible ligand for iron

Identification of the *unusual axial ligands* for iron in P-450 has been an essential problem for understanding the unique properties of P-450. Mason et al.[51] first proposed coordination of a thiolate anion to the

TABLE 3.5. Hemepeptides obtained by selective cleavage with BrCN

	Source of P-450	
	Pseudomonas putida	Liver microsomes from PB-treated rabbits
$CrSO_3H$	1	1
Asx	3	4
Thr	2	2
Ser	3	2
Glx	3	4
Pro	3	2
Gly	8	3
Ala	6	2
Val	4	2
Ile	2	2
Leu	3	4
Tyr	1	1
Phe	2	2
His	1	1
Lys	2	2
Arg	2	2
HSer	1	1
Total	47	37
Heme	1	1

(Source: ref. 48. Reproduced by kind permission of Academic Press, Inc., U. S. A.)

heme of P-450 on the basis of the finding that mercurial sulfhydryl reagents, such as p-chloromercuribenzoate (PCMB), induce the conversion of P-450 to P-420. They assumed that P-420 is the hydrate of the protoheme-protein complex, that the microsomal Fe_x named by Mason et al.[51] for the compound displaying a characteristic ESR signal is the sulfide of P-420, that P-450 in the reduced or oxidized configuration is the phospholipid complex of the sulfide of P-420, and that CO displaces sulfide but continues to interact with it and with phospholipid in the CO-complex. Although the site of action of the reagents was unknown, and the concentrations of mercurials required for the conversion (of the order of 10^{-3} M) are much greater than the concentration of the cytochrome present in the microsomes, this hypothesis has received support from data obtained by various approaches.

As mentioned above, half-cystines are found in both bacterial and microsomal P-450, and a cysteine residue is present in hemepeptides obtained by the selective chemical cleavage of P-450 with BrCN.[48] At least in bacterial P-450, all 6 of the half-cystines are present as free sulfhydryls, which can be titrated by PCMB, and the last two titratable sulfhydryl groups appear to be related to the spectrum and activity of P-450.[52] Iodine, which is known to attack sulfhydryl groups, can also induce the conversion of P-450 to P-420 under suitable conditions.[53]

Plausible observations suggesting sulfur as one of the axial ligands have been obtained by physicochemical experiments. A close similarity was revealed between the ESR spectrum of metmyoglobin-n-propyl mercaptide [54,55] and the low spin form of P-450.[24,51] Tsai et al.[56] also indicated that bacterial P-450 had an ESR signal typical of a low spin ferric-heme compound with sulfur as one of its axial ligands, based on ESR studies of a purified preparation. Recently, evidence has been accumulating from NMR [57] and MCD[58] studies to support the proposal of thiolate anion as the axial ligand. These physical properties of P-450 will be described in detail below (3.3.6 and 3.3.7).

Another approach to the problem has been to prepare a model compound having spectral characteristics similar to those of P-450. Stern and Peisach[59] showed that a complex of ferrous protoheme, CO and thiolate anion, as the 2-mercaptoethanol or n-propanethiol at alkaline pH, had absorption maxima at 450 and 555 nm in the presence of tetramethyl ammonium hydroxide. Moreover, the complex shows the same MCD pattern as P-450.[58] Other model compounds hememercaptide complexes, having an MCD or ESR spectrum similar to that of P-450 have also been reported.[58,60]

On the other hand, a number of other studies have suggested that certain hydrophobic interactions may play an important role in maintaining the integrity of P-450.[53,61] Hydrophobic bonding may well fortify the conformation of the P-450 molecule, making the sulfhydryl group accessible to the iron of the heme.

E. Oxidation-reduction potentials

Waterman and Mason[62] examined the redox potential of hepatic microsomal P-450 of PB-treated rabbits in microsomes by using a redox dyestuff. The oxidized form of P-450 remaining on setting the equilibrium potential with oxidation-reduction indicators under anaerobic conditions was measured by ESR spectroscopy. The titration curve showed an anomalous shape; in spite of an apparent midpoint at about -0.41 V, about 40% of the P-450 remained oxidized at potentials in the region of -0.44 V at pH 7.0, while the P-450 was completely reduced at more positive potentials in the presence of CO. It thus appears that more than one type of P-450 redox couple is present in hepatic microsomes, and that about 60% of the P-450 reducible with the dyestuff represents a single species with a midpoint potential of -0.34 V. Cooper et al.[63] had previously reported a potential of -0.40 V for P-450 in bovine adrenocortical mitochondria.

The redox potentials of purified preparations have recently been reported, as summarized in Table 3.6.[64,65] The presence of substrate

TABLE 3.6. Oxidation-reduction potentials of P-450

Source of P-450	Form of condition	E_0' (V) at pH 7
Liver microsomes from PB-treated rabbits	P-450	−0.33
	P-450-benzphetamine	−0.33
	P-450 (+ phospholipid)	−0.33
	P-450-benzphetamine (+ phospholipid)	−0.33
	P-450 (+ CO)	−0.15
Pseudomonas putida	P-450	−0.27
	P-450-camphor	−0.17
	P-450-camphor-O_2	0

and/or phospholipid does not significantly affect the redox potential of microsomal P-450, whereas bacterial P-450 has a more positive potential in the presence of camphor, which is ascribable to tighter binding of the substrate to the bacterial P-450 in the reduced form than in the oxidized form.[66] It is interesting, in relation to the supply of the second of the two electrons required for the overall monooxygenase reaction, that the ternary complex, P-450-substrate-O_2, which is to accept the second electron shows a much more positive potential than the substrate complex of the cytochrome. If this situation is similar to the case of microsomal P-450 (it has not yet been examined), the possibility exists that the redox potential of the ternary complex is dependent on both the molecular species of P-450 and substrates added, and determines the second electron flow from NADPH to the cytochrome either directly from NADPH-cytochrome c reductase or via cytochrome b_5.[67]

The redox potential of P-450 is thus the lowest reported for any cytochrome in its native environment. Waterman and Mason[62] suggested that the low potential may indicate a strong electron donor as a ligand for iron in P-450, such as a thiolate anion.

F. Immunochemical properties

Dus et al.[48] demonstrated certain immunochemical similarities between bacterial P-450 and hepatic microsomal P-450 from PB-treated rabbits. Hepatic microsomal P-450 exhibits 60–70% cross-reactivity with rabbit antibodies elicited against bacterial P-450, when the competition of both P-450's with ^{125}I-labeled bacterial P-450 for binding to antibacterial P-450 antibody was examined. Furthermore, antibodies against bacterial P-450 inhibit benzphetamine hydroxylation in the reconstituted microsomal enzyme system, but are not as effective as antibodies against microsomal P-450.

Using both Ouchterlony double diffusion and quantitative immunoprecipitation analysis, Thomas et al.[68-70] examined the specificity of antisera directed against hepatic microsomal P-450 from MC-treated and PB-treated rats. Antisera against P-450 from MC-treated rats can form immunoprecipitin with P-450 from both the MC- and PB-treated animals, but show a greater specificity for P-450 from MC-treated rats than for P-450 from the PB-treated animals. The antisera are about 5 times more effective in precipitating the former than the latter, judging from the ratio of the initial portions of the slopes of curves obtained by quantitative immunoprecipitation. On the other hand, antibody against P-450 from PB-treated rats reacts well with P-450 from the PB-treated animals but poorly and incompletely with P-450 from the MC-treated animals. From a detailed analysis by Ouchterlony immunodiffusion tests, it has been shown that their purified preparations contain four antigenically different forms in P-450 from PB-treated rats and two different forms in P-450 from MC-treated rats. The specificity of the antibody for inhibition of the hydroxylase activity of several substrates was also examined. Thomas *et al.* concluded that rats can synthesize at least six immunochemically distinguishable forms of the cytochrome.

3.1.3 Conversion of P-450 to P-420

On indicating the hemoprotein nature of microsomal "CO-binding pigment", Omura and Sato found that the treatment of liver microsomes with deoxycholate, snake venom phospholipase A, or Steapsin led to conversion of the pigment from P-450 to P-420, the latter possessing absorption spectra characteristic of hemoproteins.[4,14] This conversion was shown subsequently to be induced by mercurial reagents,[71] bathocuproine sulfonate,[51] urea,[51] and organic solvents.[51] Since most of the anomalous properties of P-450 disappeared on its conversion to P-420, extensive studies were made to determine the kinds of treatment that could effect the conversion, in an attempt to obtain precise information on the structural basis for the unusual spectral properties of P-450. As shown in Table 3.7, the conversion can be caused by a wide variety of treatments which seemingly have little relation to one another. However, it is possible to classify the treatments into two broad groups: (1) those which attack sulfhydryl groups, and (2) those which cause disruptions of the hydrophobic interactions.

A. Mercurials and iodine

Cooper *et al.*[71] first observed that the sulfhydryl reagent, *p*-chloromercuribenzoate (PCMB), causes conversion of P-450 to P-420 in rat liver microsomes. Other mercurials such as *p*-chloromercuriphenyl-

TABLE 3.7. Conversion of P-450 to P-420 by various reagents and methods of treatment

Reagents or treatment	Concentration	Ref.
Mercurials		
PCMB, PCMS, HgCl$_2$, mersalyl	0.1-1 mM	51, 71-73
Iodine	10 μM	53
Neutral salts	1-6 M	53
(consistent with Hofmeister's lyotropic series)		
Organic compounds		
(relative efficiency in each series of compounds increases with increase in their hydrophobic character)		
Alcohols	1-20%	51, 53, 61
Ureas	1-5 M	51, 61
Anilines	0.01-1 M	53, 61
Ketones		61
Phenols	1-100 mM	61
Amides	1-8 M	61
Detergents		
Sodium deoxycholate, sodium cholate sodium dodecyl sulfate	0.5%	4, 14, 15, 61
Lysolecithin	0.2%	53
Protein denaturants		
Urea, guanidine hydrochloride	4 M, 2 M	51, 53
Hydrolases		
Phospholipase A (heated snake venom)	0.1%	4, 14
Trypsin, Steapsin, Nagarse		4, 14, 18, 51
Bathocuproin sulfonate	1 mM	51
Acidic	pH<6	
Alkaline	pH>8	

sulfonate, HgCl$_2$, and mersalyl (sodium o-[(3-hydroxymercuri-2-methoxypropyl)carbamoyl]phenoxyacetate), are also effective, but other types of sulfhydryl reagents such as N-ethylmaleimide, iodoacetate or arsenite, show little or no effect.[51,72,73] Ichikawa and Yamano[15] have reported that mercaptide-forming reagents such as CuSO$_4$ and CdCl$_2$, can also cause the conversion.

The concentration range of mercurials over which the P-450 to P-420 conversion occurs is much greater than the concentration of P-450 present in microsomes. Moreover, the conversion proceeds slowly. For example, it has been reported to take about 20 min to induce 90% conversion of P-450 to P-420 when rabbit liver microsomes are incubated with 2.0 mM of PCMB in the presence of Lubrol W.[72] Franklin[73] has claimed that the conversion by mercurials is rather incomplete, and that the extent of conversion achieved depends on the species and prior treatment of the animals. It seems, therefore, that mercurial reagents do not act exclusively on the probable ligand sulfhydryl group of P-450, although judging from the concomitant change in spin state, the mercurial conversion may be due to a direct attack by the reagents on the ligand rather than an effect on the ligand through a conformational change in the protein molecule.[72]

When 1 mM iodine was added to a microsomal suspension containing 1–2 mg protein per ml, the P-450 was extensively destroyed, probably due to oxidative decomposition of the heme.[53] At lower concentrations of both iodine and microsomes, however, the conversion of P-450 to P-420 was observed without extensive destruction of the hemoprotein.[53] This conversion may be the result of a direct attack by iodine on the ligand itself, since the reagent can cause modifications in the protein residues, especially the sulfhydryl groups.

B. Neutral salts

Imai and Sato[53] have shown that exposure of microsomes to a high concentration of neutral salts results in conversion of P-450 to P-420. The efficiency of various salts to cause the conversion obeys the sequence known as Hofmeister's lyotropic series of ions, i.e. $SCN^- > I^- > Br^- >$

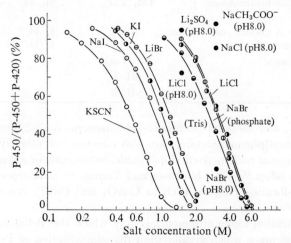

Fig. 3.1. Conversion of P-450 to P-420 as a function of the concentration of various salts added. Microsomes (2.6 mg protein per ml) in either 0.2 M potassium phosphate or Tris-acetate buffer were placed in both the sample and reference cuvettes of the spectrophotometer and incubated for 10 min aerobically in the presence of various salts. The CO-difference spectrum was then measured. The pH was 7.25 unless otherwise stated.
(Source: ref. 53 Reproduced by kind permission of Springer-Verlag, W. Germany.)

$Cl^- > SO_4^{2-}$, CH_3COO^- for anions, and $Li^+ > K^+$, Na^+ for cations, as illustrated in Fig. 3.1. The concentrations of salts required for conversion of the cytochrome are in the same range as those for destabilization of macromolecules as well as those for salting in nonelectrolytes. It appears likely that the conversion results from disturbance of the hydrophobic environment around the heme moiety, either by the primary action of the neutral salts or by secondary effects due to conformational changes in the P-450 molecule. Purified preparations of P-450 also behave similarly to the microsomal bound form toward high concentrations of lyotropic salts, so implying that the neutral salts act directly on the P-450 molecules rather than indirectly through the microsomal membranes.[32]

Microsomal P-450 from MC-treated rats is more resistant to neutral salts than that from untreated animals. This suggests that the environment of the heme moiety of P-450 is more strongly hydrophobic in MC-induced rats than in uninduced animals.[74]

The conversion induced by neutral salts is of first order and proceeds more rapidly in the reduced form of P-450 than in the oxidized form.[53] For example, when microsomes were incubated with 0.6 M NaI at pH 7.25, the conversion in dithionite-treated microsomes, in which the P-450 is kept reduced, was faster by a factor of more than 10 compared with that in aerobic microsomes. This difference in susceptibility to neutral salts between the oxidized and reduced forms may be ascribed to a conformational difference between the two forms. Yong et al.[75] have reported that ferric P-450 shows extrinsic Cotton effects, whereas these are lacking in ferrous P-450. Changes in ORD properties as a result of changes in the valence state of the heme iron probably reflect a conformational change in the environment around the heme chromophore. Alternatively, it is apparently easier to remove the thiolate ligand of the heme from its surroundings in the reduced form than in the oxidized form because the thiolate anion, serving as a natural ligand for the P-450 heme, cannot coordinate to ferrous heme. P-420 shows no extrinsic Cotton effects either in the oxidized or reduced forms.[75]

C. Organic compounds

On addition of alcohols to aerobic microsomes, spectral changes similar to those produced by drug substrates are observed.[76] However, the same solvents, at high concentrations, induce the conversion of P-450 to P-420.[51,53,61] The order of efficiency of the alcohols to induce the the conversion is isoamyl>isobutyl>isopropyl>ethyl>methyl; that is, the relative efficiency increases with increasing length of carbon chain.[53] Aniline can also induce the conversion of P-450 to P-420 at a concentration more than 10 times that for inducing Type II spectral changes.[53]

Ichikawa and Yamano[61] have examined in detail the conversion of microsomal P-450 to P-420 caused by a number of organic compounds and indicated that the conversion may depend on the hydrophobic character of the organic compounds. The transition concentrations (defined in each case as the concentration of the organic compound required for 50% conversion after incubation of the microsomes with the compound at a certain temperature for 10 min) were determined with several series of organic compounds and comparisons were made within the different series. The sequence of transition concentrations for urea derivatives thus obtained was, in increasing order, 1,3-diethyl, tetramethyl, ethyl, 1,1-dimethyl, N-methyl urea, and urea. That is to say, the transition concentration decreased with increasing number of methyl groups and increasing chain length of the alkyl groups on the nitrogen atoms of the ureas. The transition concentrations of anilines and phenols (Fig. 3.2) were closely parallel to the π values of the organic compounds, i.e. the logarithms of the distribution constants of the various organic compounds between water and 1-octanol. From these results, Ichikawa and Yamano[61] concluded that the conversion of P-450 to P-420 caused by organic compounds may be due to hydrophobic interac-

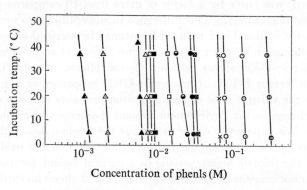

○ resorcinol   □ m-ethylphenol   ▲ 2,4,6,-trichlorophenol
○ phenol   ■ p-bromophenol   △ 2,3,4,6-tetrachlorophenol
■ p-cresol   ▫ o-iodophenol   ▲ pentachlorophenol
● m-cresol   △ 2,4-dichlorophenol   ● o-chlorophenol
◉ m-aminophenol   × p-hydroxybenzoic acid

Fig. 3.2. Conversion of P-450 to P-420 in the presence of phenols. Microsomes (2 mg protein per ml) were suspended in 0.1 M phosphate buffer, pH 7.0. Various concentrations of phenols were incubated with the suspension for 10 min at various temperatures. The abscissa shows the transition concentration.
(Source: ref. 61. Reproduced by kind permission of Elsevier Scientific Publishing Co., Netherlands.)

tions between the compounds added and the hydrophobic bonding in P-450, which may be responsible for its unusual character.

D. Detergents, protein denaturants and miscellaneous treatments

Conversion of P-450 to P-420 can be induced by ionic detergents such as sodium deoxycholate, sodium cholate, and sodium dodecyl sulfate.[4,14,61] The concentrations required for conversion correspond to those for the solubilization of P-450 from microsomes. On the other hand, in comparison with ionic detergents, P-450 is rather stable against conversion to P-420 by non-ionic detergents such as Emulgens, in spite of their good solubilization ability.[21,26] Typical protein-denaturing reagents such as urea[51] and guanidine hydrochloride[53] can also induce the conversion of P-450 to P-420. The concentrations required for complete conversion are 4 M and 2 M in the case of urea and guanidine hydrochloride, respectively, at neutral pH. The action of such detergents and protein denaturants on P-450 can be explained as a result of the direct and/or indirect destruction of the hydrophobic environment around the heme moiety of P-450, since both kinds of reagents can disrupt the hydrophobic bonding and destroy the native conformation of protein molecules.

Anaerobic treatment of microsomes with proteases such as trypsin results in the conversion of P-450 to P-420, as mentioned above. Proteases primarily cause modifications of the native conformation of the cytochrome, breaking down secondarily the hydrophobic environment around the heme. When microsomes were treated with phospholipase A, P-450 was converted to P-420.[4,14] Since phospholipase A produces lysolecithin from lecithin, and liver microsomes contain large amounts of lecithin, it appeared likely that the conversion induced by phospholipase A was actually due to the formation by this enzyme of lysolecithin which has powerful detergent activity. In fact, anaerobic incubation of microsomes with 0.2% lysolecithin caused complete conversion of P-450 to P-420.[53] On the other hand, phospholipase C, which splits phosphoryl choline from lecithin, caused only 20% conversion, although this treatment removed 50% of the total phopholipid phosphorus from the microsomes.[53]

Certain chelating agents such as bathocuproine sulfonate,[51] which can induce the conversion, are soluble in both aqueous and lipid solvents and may also act as detergents.

P-450 is stable only in the neutral pH region between 6 and 8. At pH values lower than 6 and higher than 8, it is converted to P-420 to some extent. In the presence of the lyotropic salts described above, P-450 is most stable at around pH 7.25, and the stability declines rapidly on both sides of the above pH region.[53] This profile appears to reflect the effect of pH on the stability of the ligand state of P-450, due either to some direct action on the ligand itself or to some indirect action on the ligand through the protein conformation.

E. Reconversion of P-420 to P-450

As mentioned above, glycerol and other polyols prevent the conversion of P-450 to P-420 on treatment with detergents or proteases.[15,18] Moreover, the P-420 which is produced by cholate or mercurials can be converted back to P-450 by treatment with polyols, glutathione or cysteine.[15] Ichikawa and Yamano[15] have reported that PCMB-produced P-420 can be converted back to P-450 almost completely with reduced glutathione. Franklin, however, has claimed that only a small reconversion can be achieved with reduced glutathione after mersaryl treatment.[73]

F. Multiplicity of P-420

Based on ESR studies of P-450, Mason et al.[51,72] suggested that the conversion of P-450 to P-420 involves passage through several states. The optically measurable conversions of P-450 to P-420 by mercurials are accompanied by a quantitative change in the spin state of the heme iron.[72] When microsomes are treated with the mercurials, the low spin signal at $g = 2.25$ characteristic of P-450 disappears, and simultaneously, a high spin signal at $g = 6.1$ appears. On the other hand, the low spin state still remains despite the considerable conversion of P-450 to P-420 when the microsomes are treated with certain other reagents, such as butanol, sodium cholate, or trypsin.[15,51,61] No spin signal is observed after treatment with sodium dodecyl sulfate, urea, or certain other reagents.[15,51,61] The P-420 produced by cholate can be converted back to P-450 by treatment with polyols or reduced glutathione, as described above, but the P-420 obtained by treatment with sodium dodecyl sulfate cannot be reconverted to P-450.[15] Moreover, P-420 which has been converted from purified P-450 by KSCN treatment exhibits absorption spectra different from those of partially purified P-420, although the CO-compounds of both P-420's possess an identical spectrum.[32,58] These observations can be reasonably explained by regarding so-called P-420 as comprehensively including several conformational states of the cytochrome, since P-420 was originally designated as a denatured product of P-450 characterized by a spectral change exhibiting a peak at 420 nm on combination with CO.[4,14]

When purified P-450 is treated with a suitable concentration of mercurials, KSCN, sodium deoxycholate, or guanidine hydrochloride in the presence of glycerol, conversion to P-420 occurs[34] but the extent of the conformational change in the protein moiety is not the same in all cases. Judging from the CD spectra in the far ultraviolet region, the helical content of the protein was hardly altered on conversion with deoxycholate, while it was strongly diminished in the case of guanidine hydrochloride.[7] Thus, conversion from P-450 to P-420 is not necessarily accompanied by a change in the gross helical conformation of the protein moiety.

The conversion of P-450 to P-420 implies replacement of the natural ligand for P-450 heme iron with another group. Recently, several lines of evidence have accumulated to support the hypothesis that one of the axial ligands for the iron in the P-450 heme is thiolate anion. On the other hand, it appears that hydrophobic bonding in P-450 also plays an important role in maintaining the structure of the cytochrome which is responsible for its unique properties, as discussed above. In summary, the following mechanism for the conversion may be proposed. When the axial ligand for the heme iron of P-450 (probably thiolate anion) is removed and replaced by a residue that can coordinate to a heme iron (amino acid residues of the protein moiety, water, or reagents added) for some particular reason, the unique properties of P-450 are replaced by those usual for a hemoprotein having protoheme as a prosthetic group. This replacement is observable spectrophotometrically as a conversion of the P-450 to P-420. Thus, the conversion occurs when the natural ligand is destroyed; this represents the case where the reagents directly attack the ligand of P-450 heme itself. When the natural ligand is not destroyed chemically, it is necessary to effect the conversion in such a way that the natural ligand is pulled apart from the heme iron to induce spatial reorganization of the ligands. This may be achieved by destruction of the hydrophobic bonding which fortifies the environment of the heme, or by some change in the native conformation of the P-450 molecule.

In connection with the involvement of the hydrophobic environment in the integrity of the unique character of P-450, it is interesting to note that a similar situation is also observed in another atypical cytochrome of the *c* type from *Rhodospirillum rubrum*, namely cytochrome *cc'*.[77] This cytochrome shows quite unusual absorption spectra at neutral pH values, whereas the spectra become characteristic of the *c*-type cytochromes at pH 12.5.[78] Conversion of the atypical form to the typical form can be induced not only by a pH change but also by various treatments, such as addition of organic solvents or urea, which are known to disrupt hydrophobic interactions.[77]

G. Interconversion between Bacterial P-450 and P-420

Yu et al.[52] have described the conversion of bacterial P-450 to P-420 by urea, guanidine hydrochloride, acetone, PCMB, and deoxycholate. The conversion rates depend on the reagent concentrations. Addition of the substrate, camphor, protects the P-450 from denaturation by deoxycholate, while it exerts only a slight effect in the case of urea. The reconversion of freshly prepared P-450 to P-420 can be induced by sulfhydryl compounds. Cysteine is the most effective reagent, although the reconversion depends on the concentration of the reagent added. Homocysteine, glutathione, dithioerythritol, β-mercaptethanol, thioglycolic acid, and mercaptosuccinic acid are less effective and results obtained with these reagents are variable. Success in reconversion also depends on the method used to produce the P-420. For example, treatment with $>25\%$ acetone at 25°C gave a product not reversible to P-450, whereas the P-420 formed with 23% acetone at room temperature could be reconverted to P-450 by the addition of 5% cysteine after removal of the acetone. More than 90% of the original activity was regained after 3 h incubation at room temperature. All of the 6 sulfhydryl groups present in native P-450 are titratable with PCMB. First, 4 of the sulfhydryl groups react rapidly with PCMB, whether camphor is present or not, and this is followed by a much slower reaction with the 2 remaining groups. During the first period, very little of the P-450 activity disappears, and the level of P-420 remains low. Glycerol or polyols cannot induce the reconversion of bacterial P-420 to P-450.

3.1.4 Interaction of P-450 with Isocyanides

A. Reduced form

In order to obtain evidence that their CO-binding pigment was a hemoprotein, Omura and Sato[4] examined the interaction between ferrous P-450 and ethyl isocyanide, a reagent known to combine with hemoglobin. At pH 7.0, where the interaction was tested, a difference spectrum characteristic of hemoproteins was obtained, although a small anomalous peak at 455 nm (the "455 nm peak") was obtained in addition to the main peak at 430 nm (the "430 nm peak") in the Soret region.[4] Imai and Sato[79] later examined in detail the properties of the ethyl isocyanide-difference spectrum and found that the relative magnitudes of the two peaks depend on the pH. The height of the 455 nm peak increases progressively with increasing pH, and this is accompanied by a corresponding decrease in the height of the 430 nm peak (see Fig. 3.3). Both forms have the same affinity for ethyl iso-

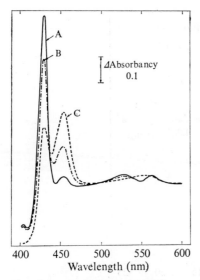

Fig. 3.3. Ethyl isocyanide-difference spectra of dithionite-treated liver microsomes measured at different pH's.[79] Microsomes (2.4 mg protein per ml) suspended in 0.2 M potassium phosphate buffer were used. The final concentration of ethyl isocyanide was 3.0 mM. A, pH 6.0; B, pH 7.0; C, pH 8.0.

cyanide. Moreover, both peaks are competitively affected by CO to the same extent at all pH values and ethyl isocyanide concentrations tested. It can thus be said that the ethyl isocyanide compound of microsomal P-450 exists in two interconvertible forms which are in a pH-dependent equilibrium. This equilibrium is affected not only by the pH but also by the ionic strength.[80] A low concentration of certain alcohols, which is lower than that required for spectral change induction in the oxidized form, also affects the 455 to 430 nm ratio in the ethyl isocyanide-difference spectrum of microsomes without conversion of P-450 to P-420, and the efficiency of the alcohols parallels their chain length (see Fig. 3.4).[81] The addition of Emulgen 913 also leads to a significant increase in the 455 nm form and a corresponding decrease in the 430 nm form, when examined with purified P-450.[34] Since pH- and ionic strength-dependent changes in the two Soret peaks of the absolute spectrum of the ethyl isocyanide compound of the reduced form are observed with purified P-450 (Fig. 3.5), it can be said that these represent inherent properties of P-450 and not apparent ones because of its binding to the microsomal membranes.[32,34] The 430 nm peak in the spectrum of the ethyl isocyanide compound of ferrous P-450 is not due to P-420 formation, since the spectrum of the ethyl isocyanide compound of ferrous P-420 exhibits a peak at 434 nm in the Soret region which

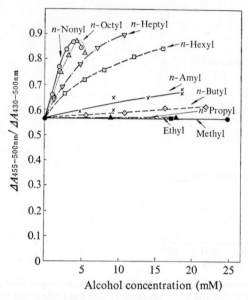

Fig. 3.4. Effect of side-chain length of alcohols on the ethyl isocyanide-difference spectrum of ferrous P-450. The difference spectra occurring on addition of 0.2 mM ethyl isocyanide to dithionite-treated microsomes (1.36 mg protein per ml) suspended in 0.2 M potassium phosphate buffer, pH 7.5, were recorded in the presence and absence of various alcohols. The ordinate shows the ratio of absorbance increments in the spectra.

is clearly different from that of the 430 nm state of the ethyl isocyanide compound of ferrous P-450.[32]

Ichikawa and Yamano[82] have reported the results of spectrophotometric and magnetic studies on compounds of P-450 with various isocyanides; methyl, ethyl, *tert*-butyl and phenyl isocyanide. All of the compounds of ferrous P-450 exhibit two peaks at 455 and 430 nm in the Soret region, and both peaks are affected by the pH in a similar way to that observed for the ethyl isocyanide interaction. However, the spectra observed at a particular pH vary with the isocyanide species. The 455 nm peak increases with increasing carbon number of the isocyanides, and this is accompanied by a corresponding decrease in the height of the 430 nm peak. The shape of the isocyanide dissociation curves of P-450 conforms to the theoretical curves for $n = 1.0$, and the dissociation constants of the compounds of P-450 with the various isocyanides are essentially identical. The values fall within the order of 10^{-6} M and are little affected by the pH.

Liver microsomes from MC-treated rats show a higher 455 to 430 nm ratio in the ethyl isocyanide-difference spectrum of ferrous P-450

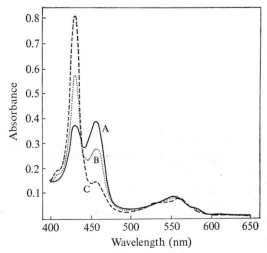

Fig. 3.5. Absolute spectrum of ethyl isocyanide compound of ferrous P-450.[32] A partially purified preparation of P-450 was dissolved in 0.2 M potassium phosphate buffer of indicated pH (A, 8.0; B, 7.0; C, 6.0;) containing 20% of glycerol and 0.8 mM of ethyl isocyanide. This final concentration of P-450 was 6.46 μM. The solution was treated with a few mg of sodium dithionite prior to measurement of the spectrum.

than do those from untreated animals (Fig. 3.6).[12,13] On the other hand, the administration of PB causes a small increase in the 455 to 430 nm ratio.[74] It has been confirmed in purified preparations that the major P-450 species from MC-treated animals shows a higher 455 to 430 nm ratio than that from PB-treated animals.[83] The position of the 455 nm peak is shifted to 453 nm in MC-treated animals, while there is no observable shift in the 430 nm peak. Bacterial P-450 gives only a peak at 453 nm.[84]

B. Oxidized form

Nishibayashi et al.[85] were the first to report that the addition of ethyl isocyanide to aerobic microsomes also induces spectral changes which show a peak at 434 nm in the Soret region, and Ichikawa and Yamano[82] have examined in detail the interaction with various isocyanides. The shape of the difference spectrum is not affected by the pH or the species of isocyanides in the oxidized form, but the dissociation constant decreases with increasing carbon number of the isocyanides. Thus, the values fall within the order of 10^{-3}, 10^{-4}, 10^{-5} and 10^{-6} for the methyl, ethyl, tert-butyl and phenyl derivatives, respectively. This suggests that the hydrophobic character of the environment around the heme moiety is strong in the oxidized form. The shape of the isocyanide

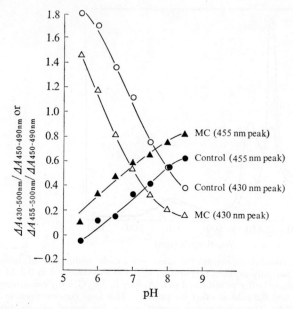

Fig. 3.6. Dependence of the 455 and 430 nm peak of the ethyl isocyanide-difference spectra on pH. Treated rats received MC (2 mg/100 g body wt, i.p.) daily for 3 days. Microsomes in 0.2 M potassium phosphate buffer and 30% glycerol were used. The final concentration of ethyl isocyanide was 0.4 mM. $\Delta A_{450-490nm}$, $\Delta A_{430-500nm}$ $\Delta A_{455-500nm}$ were taken as the CO-difference spectrum, 430 and 455 nm peaks of the ethyl isocyanide-difference spectrum, respectively.
(Source: ref. 74. Reproduced by kind permission of Academic Press, Inc., U.S.A.)

dissociation curves for ferric P-450 also conforms to the theoretical curves for $n = 1.0$. The spectral changes induced by saturated concentrations of the isocyanides are of the same magnitude irrespective of the isocyanide derivative. The addition of ethyl isocyanide to the oxidized form of purified P-450 causes the same spectral changes as in the case of aerobic microsomes.[34]

C. Effects of isocyanide on the oxidation-reduction cycle

Imai and Sato[86] have examined the spectral change induced in liver microsomes by ethyl isocyanide in the presence of NADPH under aerobic conditions. The resultant spectrum cannot be explained by a simple summation of the ethyl isocyanide-difference spectra of oxidized and dithionite-reduced microsomes. Based on spectral analyses, it was found that the steady state levels of the 455 nm and 430 nm states of the isocyanide compound of ferrous P-450 are different in the presence of both NADPH and atmospheric oxygen from the level in dithionite-

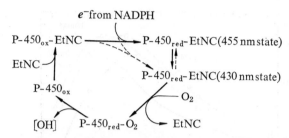

Fig. 3.7. The oxidation-reduction cycle which P-450 undergoes in the presence of NADPH, air and ethyl isocyanide (EtNC).[86]

reduced material; the 455 nm state is definitely predominant over the 430 nm state. The higher reduction level in the 455 nm state than in the 430 nm state implies (1) that the competition of the isocyanide with oxygen for ferrous P-450 takes place in the 430 nm state rather than in the 455 nm state, and (2) that the conversion of the 455 nm state to the 430 nm state is slower than the supply of the former by NADPH-linked reduction of the oxidized form (Fig. 3.7). Thus, the reduction of microsomal P-450 by NADPH seems to produce preferentially the 455 nm state of the hemoprotein. The ethyl isocyanide bound to the heme appears to be replaced rapidly by oxygen, when the 455 nm state is converted to the 430 nm state.

Imai and Sato[87] have described the effect of ethyl isocyanide on the aniline hydroxylase activity of rabbit liver microsomes. Despite the competition of ethyl isocyanide with oxygen for ferrous P-450, the activity under standard assay conditions was activated, rather than inhibited, by 10 μM to 1 mM ethyl isocyanide (Fig. 3.8). However, when the ethyl isocyanide concentration was increased further, a gradual decrease in the degree of activation was observed. This dependency of the activity on the ethyl isocyanide concentration suggests that the reagent has both stimulatory and inhibitory effects on the hydroxylase system. From the effect of oxygen tension on the activity in the presence of ethyl isocyanide, it was concluded that the inhibitory effect is due to competition of the reagent with oxygen for the heme of ferrous P-450. The stimulatory effect, which is visualized exclusively by extrapolating the activity at infinite oxygen tension, was analyzed by changing the ethyl isocyanide concentration, aniline concentration and pH. It is suggested that the stimulatory effect results from a change in the reactivity of ferric P-450 caused by its combination with ethyl isocyanide, and it is affected by the pH. At pH 6.5 only the inhibitory effect is observable, but the stimulatory effect emerges and increases progressively as the pH is increased up to 8.0. The contributions of these two

68 MOLECULAR PROPERTIES

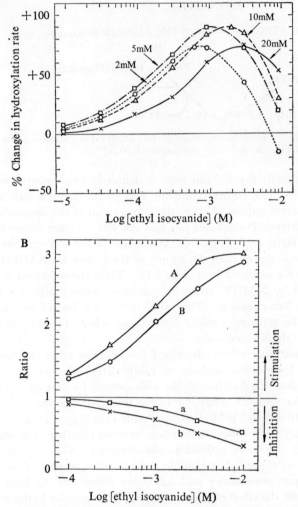

Fig. 3.8. Effects of ethyl isocyanide on aniline hydroxylation.[87] A, *Activity in air*: The standard assay conditions were used, except that the aniline concentration was varied as indicated and ethyl isocyanide was added at the indicated concentrations. B, *Stimulatory and inhibitory effects*: The ideal activity was calculated from Lineweaver-Burk plots of the dependence of activity on oxygen tension from the intercepts on the ordinate as the activity at infinite oxygen tension.

Curves A and B: effect of ethyl isocyanide concentration on the ratio [ideal activity (+ ethyl isocyanide)]/[ideal activity (− ethyl isocyanide)]. The aniline concentrations for curves A and B were 10 mM and 20 mM, respectively.

Curves a and b: effect of ethyl isocyanide on the ratio [activity (measured)]/[activity ($O_2 \to \infty$)]. The oxygen concentrations for curves a and b were 20% and 10%, respectively.

effects also differ depending on the source of the microsomes and the nature of the hydroxylatable substrate.

D. Relationship between the 455 and 430 nm states

A number of studies have suggested the importance of certain hydrophobic interactions in maintaining the integrity of P-450. The observations described in this section indicate that the 455 nm state of the isocyanide compound increases, and the 430 nm state correspondingly decreases, when the hydrophobicity of the enviroment around the heme moiety of P-450 is increased, suggesting that the 455 nm state, rather than the 430 nm state, should be ascribed to the hydrophobic character of the environment around the heme moiety. This idea is also supported by the spectral behavior of the protoheme-ethyl isocyanide system, a model system for P-450 isocyanide interactions. Imai and Sato[88] have reported that at neutral pH values, dithionite-reduced protoheme interacts with ethyl isocyanide to form three spectrally different compounds; a mono-isocyanide compound, a normal di-isocyanide compound and an anomalous compound, which have Soret peaks at 414, 428 and 455 nm, respectively. Among them, the mono-isocyanide compound exists only when the concentration of ethyl isocyanide is insufficient to produce the di-isocyanide compound completely. A decrease in the 455 nm form of the isocyanide compound of protoheme and a corresponding increase in the 428 nm form are observed on various treatments that disrupt hydrophobic interactions, such as the addition of organic solvents. On the other hand, the 455 nm form increases under conditions which favor the formation of aggregates probably by hydrophobic interactions.

Although it is unknown whether the 5th ligand for the iron in the 455 nm state of isocyanide compounds of ferrous P-450 is different from that in the 430 nm state or not, there may be a certain conformational difference at least in the environment around the heme chromophore of P-450 between these two states of the isocyanide compounds. This is suggested by the following observations. (1) When the CO-compound of P-450 is irradiated with 450 nm monochromatic light at temperatures below 0°C in the presence of ethyl isocyanide, photo-dissociation of the CO-compound and a corresponding formation of the ethyl isocyanide compound are observed (Fig. 3.9)[89]. The relative rates of formation of the two ethyl isocyanide compounds indicate that on replacenment of the CO by ethyl isocyanide as the ligand for the heme iron of P-450, the 455 nm state of the isocyanide compound is formed first and is then converted to the 430 nm state. This suggests that the conversion of the 455 nm form to the 430 nm form proceeds at a slower rate than the re-

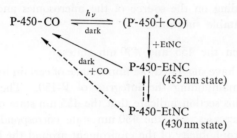

Fig. 3.9. Reaction occurring on photo-irradiation of the CO-compound of P-450 in the presence of ethyl isocyanide (EtNC).

placement of CO by ethyl isocyanide. (2) In the steady state established in the presence of NADPH, oxygen and ethyl isocyanide at room temperature, the ethyl isocyanide compound of ferrous P-450 exists mainly in the 455 nm state rather than in the 430 nm state. This suggests that the conversion of the 455 nm to the 430 nm state is slower than the supply of the former by NADPH-linked reduction of ferric P-450 and the replacement of the ligand, ethyl isocyanide of the 430 nm state by oxygen, followed by oxidation of the ferrous to ferric P-450.[86] (3) Circular dichroism measurements indicate that, optically speaking, the 455 nm state is much more asymmetric than the 430 nm state. This suggests that a considerable difference exists in the geometries of the environments around the heme chromophore between the two forms.[21] (4) The following has been suggested from studies of the MCD of the ethyl isocyanide compound of ferrous P-450.[58] The heme in the 455 nm form is in a state resembling that in the CO-compound of ferrous P-450, while the 430 nm form may constitute a state in which thiolate anion is replaced by ethyl isocyanide which coordinates to the heme *trans* to a strong ligand such as the imidazole nitrogen.

References

Section 3.1
1. M. Klingenberg, *Arch. Biochem. Biophys.*, **75**, 376 (1958).
2. D. Garfinkel, *ibid.*, **77**, 493 (1958).
3. T. Omura and R. Sato, *J. Biol. Chem.*, **237**, PC1375 (1962).
4. T. Omura and R. Sato, *ibid.*, **239**, 2370 (1964).
5. Y. Imai and R. Sato, *Biochem. Biophys. Res. Commun.*, **60**, 8 (1974).
6. C. Hashimoto and Y. Imai, *ibid.*, **68**, 821 (1976).
7. H. Satake, Y. Imai, C. Hashimoto, R. Sato, T. Shimizu, Y. Nozawa and M. Hatano, *Seikagaku* (Japanese), **48**, 508 (1976).
8. T. A. van der Hoeven, D. A. Haugen and M. J. Coon, *Biochem. Biophys. Res. Commun.*, **60**, 569 (1974).

9. D. A. Haugen and M. J. Coon, *J. Biol. Chem.*, **251**, 7929 (1976).
10. D. Ryan, A. Y. H. Lu, J. Kawalek, S. B. West and W. Levin, *Biochem. Biophys. Res. Commun.*, **64**, 1134 (1975).
11. J. C. Kawalek, W. Levin, D. Ryan, P. E. Thomas and A. Y. H. Lu, *Mol. Pharmacol.*, **11**, 874 (1975).
12. A. P. Alvares, G. Schilling, W. Levin and R. Kuntzman, *ibid.*, **29**, 512 (1968).
13. N. E. Sladek and G. J. Mannering, *ibid.*, **24**, 668 (1966).
14. T. Omura and R. Sato, *J. Biol. Chem.*, **239**, 2379 (1964).
15. Y. Ichikawa and T. Yamano, *Biochim. Biophys. Acta*, **131**, 490 (1967).
16. A. Y. H. Lu and M. J. Coon, *J. Biol. Chem.*, **243**, 1332 (1968).
17. A. Y. H. Lu, R. Kuntzman, S. West and A. H. Conney, *Biochem. Biophys., Res. Commun.*, **42**, 1200 (1971).
18. H. Nishibayashi and R. Sato, *J. Biochem. (Tokyo)*, **63**, 766 (1969).
19. Y. Ichikawa and T. Yamano, *Biochim. Biophys. Acta*, **200**, 220 (1970).
20. F. Mitani, A. P. Alvares, S. Sassa and A. Kappas, *Mol. Pharmacol.*, **7**, 280 (1971).
21. H. Satake and Y. Imai, *Seikagaku* (Japanese), **43**, 208 (1971).
22. K. Comai and J. L. Gaylor, *J. Biol. Chem.*, **248**, 4947 (1973).
23. A. Y. H. Lu and W. Levin, *Biochim. Biophys. Acta*, **344**, 205 (1974).
24. Y. Miyake, J. L. Gaylor and H. S. Mason, *J. Biol. Chem.*, **243**, 5788 (1968).
25. T. Fujita, D. W. Shoeman and G. J. Mannering, *ibid.* **248**, 2192 (1973).
26. R. Sato, H. Satake and Y. Imai, *Drug Metab. Disposition*, **1**, 6 (1973).
27. A. Y. H. Lu and W. Levin, *Biochem. Biophys. Res. Commun.*, **46**, 1334 (1972).
28. W. Levin, A. Y. H. Lu, D. Ryan, S. West, R. Kuntzman and A. H. Conney, *Arch. Biochem. Biophys.*, **153**, 543 (1972).
29. A. Y. H. Lu, S. B. West, D. Ryan and W. Levin, *Drug Metab. Disposition*, **1**, 29 (1973).
30. W. Levin, D. Ryan, S. West and A. Y. H. Lu, *J. Biol. Chem.*, **249**, 1747 (1974).
31. T. A. van der Hoeven and M. J. Coon, *ibid.*, **249**, 6302 (1974).
32. Y. Imai and R. Sato, *J. Biochem. (Tokyo)*, **75**, 689 (1974).
33. Y. Imai, *ibid.*, **80**, 267 (1976).
34. Y. Imai, *unpublished data.*
35. M. D. Maines and M. W. Anders, *Arch. Biochem. Biophys.*, **159**, 201 (1973).
36. U. Muller-Eberhard, H. H. Liem, C. A. Yu and I. C. Gunsalus, *Biochem. Biophys. Res. Commun.*, **35**, 229 (1969).
37. C. A. Yu and I. C. Gunsalus, *J. Biol. Chem.*, **249**, 107 (1974).
38. T. Kamataki, D. M. Belcher and R. A. Neal, *Mol. Pharmacol.*, **12**, 921 (1976).
39. M. T. Huang, S. B. West and A. Y. H. Lu, *J. Biol. Chem.*, **251**, 4659 (1976).
40. C. A. Yu, I. C. Gunsalus, M. Katagiri, K. Suhara and S. Takemori, *ibid.*, **249**, 94 (1974).
41. K. Dus, M. Katagiri, C. A. Yu, D. L. Erbes and I. C. Gunsalus, *Biochem. Biophys. Res. Commun.*, **40**, 1423 (1970).
42. S. Takemori, H. Sato, T. Gomi, K. Suhara and M. Katagiri, *ibid.*, **67**, 1151 (1975).
43. S. Takemori, K. Suhara, S. Hashimoto, M. Hashimoto, H. Sato, T. Gomi and M. Katagiri, *ibid.*, **63**, 588 (1975).
44. A. Dietch and R. Sato, *unpublished data.*
45. R. A. Capaldi and G. Vanderkooi, *Proc. Natl. Acad. Sci. U.S.A.*, **69**, 930 (1972).
46. M. Shikita and P. F. Hall, *J. Biol. Chem.*, **248**, 5598 (1973).
47. M. Shikita and P. F. Hall, *ibid.*, **248**, 5605 (1973).
48. K. Dus, W. J. Litchfield, A. G. Miguel, T. A. Van der Hoeven, W. L. Dean and M. J. Coon, *Biochem. Biophys. Res. Commun.*, **60**, 15 (1974).
49. K. Dus, R. Goewert, C. L. Weaver, D. Carey and C. A. Appleby, *ibid.*, **69**, 437 (1976).
50. D. A. Haugen, L. G. Armes, K. T. Yasunobu and M. J. Coon, *ibid.*, **77**, 967 (1977).
51. H. S. Mason, J. C. North and M. Vanneste, *ibid.*, **24**, 1072 (1965).
52. C. A. Yu and I. C. Gunsalus *J. Biol. Chem.*, **249**, 102 (1974).

53. Y. Imai and R. Sato, *Eur. J. Biochem.*, **1**, 419 (1967).
54. C. R. E. Jefcoate and J. L. Gaylor, *Biochemistry*, **8**, 3464 (1969).
55. von A. Röder and E. Bayer, *Eur. J. Biochem.*, **11**, 89 (1969).
56. R. Tsai, C. A. Yu, I. C. Gunsalus, J. Peisach, W. Blumberg, W. H. Orme-Johnson and H. Beinert, *Proc. Natl. Acad. Sci., U.S.A.*, **66**, 1157 (1970).
57. I. Morishima, T. Iizuka and Y. Ishimura, *Seikagaku* (Japanese), **46**, 535 (1974).
58. T. Shimizu, T. Nozawa, M. Hatano, Y. Imai and R. Sato, *Biochemistry*, **14**, 4172 (1975).
59. J. O. Stern and J. Peisach, *J. Biol. Chem.*, **249**, 7495 (1974).
60. J. P. Collman, T. N. Sorrell and B. M. Hoffman, *J. Am. Chem. Soc.*, **97**, 913 (1975).
61. Y. Ichikawa and T. Yamano, *Biochim. Biophys. Acta*, **147**, 518 (1967).
62. M. R. Waterman and H. S. Mason, *Arch. Biochem. Biophys.*, **150**, 57 (1972).
63. D. Y. Cooper, H. Shleyer and O. Rosenthal, *Ann. N. Y. Acad. Sci.*, **174**, 205 (1970).
64. F. P. Guengerich, D. P. Ballou and M. J. Coon, *J. Biol. Chem.*, **250** 7405 (1975).
65. K. Dus, *The Enzymes of Biological Membranes* (ed. A. Martonosi), IV, p. 199, Plenum Press (1976).
66. B. W. Griffin and J. A. Peterson, *Biochemistry*, **11**, 4740 (1972).
67. Y. Imai and R. Sato, *Biochem. Biophys. Res. Commun.*, **75**, 420 (1977).
68. P. E. Thomas, A. Y. H. Lu, D. Ryan, S. West and W. Levin, *Life Sci.*, **15**, 1475 (1974).
69. P. E. Thomas, A. Y. H. Lu, D. Ryan, S. B. West, J. Kawalek and W. Levin, *J. Biol. Chem.*, **251**, 1385 (1976).
70. P. E. Thomas, A. Y. H. Lu, D. Ryan, S. B. West, J. Kawalek and W. Levin, *Mol. Pharmacol.*, **12**, 746 (1976).
71. D. Y. Cooper, S. Narasimhulu, O. Rosenthal and R. W. Estabrook, *Oxidase and Related Redox Systems* (ed. T. King, H. S. Mason and M. Morrison), p.838, John Wiley (1965).
72. K. Murakami and H. S. Mason, *J. Biol. Chem.*, **242**, 1102 (1967).
73. M. R. Franklin, *Mol. Pharmacol.*, **8**, 711 (1972).
74. Y. Imai and P. Siekevitz, *Arch. Biochem. Biophys.*, **144**, 143 (1971).
75. F. C. Yong, T. E. King, S. Oldham, M. R. Waterman and H. S. Mason, *ibid.*, **138**, 96 (1970).
76. Y. Imai and R. Sato, *J. Biochem. (Tokyo)*, **62**, 239 (1967).
77. Y. Imai, K. Imai, R. Sato and T. Horio, *ibid.*, **65**, 225 (1969).
78. T. Horio and M. Kamen, *Biochim. Biophys. Acta*, **48**, 266 (1961).
79. Y. Imai and R. Sato, *J. Biochem. (Tokyo)*, **62**, 464 (1967).
80. Y. Imai and R. Sato, *ibid.*, **63**, 270 (1968).
81. Y. Imai and H. S. Mason, *unpublished data*.
82. Y. Ichikawa and T. Yamano, *Biochim. Biophys. Acta*, **153**, 753 (1968).
83. C. Hashimoto, Y. Imai and R. Sato, *Seikagaku* (Japanese), **48**, 508 (1976).
84. B. Griffin and J. A. Peterson, *Arch. Biochem. Biophys.*, **145**, 220 (1971).
85. H. Nishibayashi, T. Omura and R. Sato, *Biochim. Biophys. Acta*, **118**, 651 (1966).
86. Y. Imai and R. Sato, *J. Biochem. (Tokyo)*, **63**, 370 (1968).
87. Y. Imai and R. Sato, *ibid.*, **63**, 380 (1968).
88. Y. Imai and R. Sato, *ibid.*, **64**, 147 (1968).
89. Y. Imai and H. S. Mason, *J. Biol. Chem.*, **246**, 5970 (1971).

3.2 Optical Properties

3.2.1 Development of Research

A. Spectral studies immediately after the discovery of microsomal "CO-binding pigment"

Since P-450 was first found as a membrane-bound "CO-binding pigment" of liver microsomes, the early spectral studies on this pigment were restricted to studies of difference spectra using liver microsomes without any further treatment. Although this pigment was later proved to be a hemoprotein, its absorption spectra are quite "abnormal" compared with those of usual hemoproteins. In the early stages of research, therefore, enough evidence to conclude that the pigment was a hemoprotein could not be derived from difference spectra, i.e. the difference between two kinds of absolute spectra. However, it must be emphasized that many of the basically important properties of this pigment were found by the early authors using the difference spectrum technique.

In 1958, three years after its discovery by G. R. Williams, Klingenberg[1] described that the CO-binding pigment showing an absorption

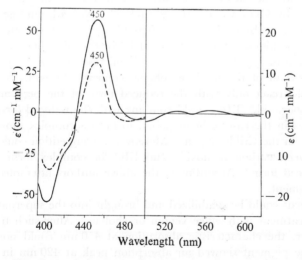

Fig. 3.10. Carbon monoxide-difference spectra of rat liver microsomes. The millimolar extinction coefficients refer to the cytochrome b_5 present in the microsomes. - - -, Carbon monoxide with NADH reduction; ———, carbon monoxide with dithionite reduction.

(Source: ref. 1. Reproduced by kind permission of Academic Press. Inc., U.S.A.)

maximum at 450 nm could be reduced by treatment of rat liver microsomes with either NADH or dithionite, that the pigment was different from cytochrome b_5, and that its absorption disappeared almost completely after treatment with cholate (Fig. 3.10). He also found small absorption peaks which were distinguishable from those of b_5 in both the visible and Soret regions of the reduced minus oxidized difference spectrum.

Garfinkel[2] showed not only that the pigment was present in liver microsomes of the pig and many other animals but that it was also decomposed by exposure to low pH's, by treatment with proteases, cholate, or digitonin. From these observations, he suggested the possible relevance of microsomal lipids to the structure of the pigment.

The hemoprotein nature of the carbon monoxide-binding pigment could not be elucidated by these studies and the pigment was assumed to be a metalloprotein.[2] This was based mainly on the fact that, in contrast to usual hemoproteins, no characteristic α- and β-peaks could be detected in the visible region, and also that photo-dissociation of the carbon monoxide complex could not be demonstrated at the illumination intensities used in those days.[1]

B. Demonstration of the pigment as a hemoprotein

The hemoprotein nature of the carbon monoxide-binding pigment was first shown by Omura and Sato[3-5] in 1962. Notwithstanding the abnormal behavior observed in the difference spectra, these authors suspected that the pigment was a hemoprotein for the following reasons. First, the amount of cytochrome b_5 in hepatic microsomes could account for only about half of the total protoheme content. Second, carbon monoxide could react only with the reduced form of the pigment. Third, ethyl isocyanide (EIC) could combine with the reduced form of the pigment, as in the case of hemoglobin, to form a complex showing α- and β-peaks in the visible region. Moreover, a competition could be observed between carbon monoxide and EIC for complex formation with the reduced form. Accordingly, the above authors attempted to isolate the pigment.

The pigment could be solubilized and brought into the supernatant fraction by treatment with snake venom or cholate. After such treatment, however, the characteristic absorption at 450 nm could not be found, and the pigment showed an absorption peak at 420 nm in the CO-difference spectrum. The membrane-bound pigment showing the peak at 450 nm was named P-450, and the solubilized pigment showing the peak at 420 nm was named P-420 by these authors. In the conversion of P-450 to P-420, P-420 was unstable and decomposed under aerobic conditions, but was stable under anaerobic conditions. The increase in absorption at 420 nm paralleled the decrease in absorption at 450 nm when the conversion was carried out anaerobically. After solubilization and purification, P-420 was found to be stable unless dithionite was added in the presence of oxygen. Since the cytochrome b_5 in microsomes could be removed by digestion with steapsin, Omura and Sato extracted P-420 from the residue by digestion with snake venom, and purified it. They found P-420 to be a hemoprotein showing typical protoheme-hemochrome type absorption spectra.

P-450 was thus proved by indirect means to be a hemoprotein, using P-420 as the mediator of verification. Furthermore, by using the CO-difference spectrum and the molar extinction coefficient increment determined by Omura and Sato, quantitative estimation of P-450 became possible. Such estimations then contributed to the development of studies on the inductive synthesis of microsomal mixed-function oxidase systems. In addition to the above results, other properties of P-450 were described, viz. that microsomal P-450 could be reduced by NADPH under anaerobic conditions,[2-4] whereas cytochrome b_5 could be reduced by NADH even under aerobic conditions, and that P-450 could combine with EIC[4] as well as carbon monoxide. It was still difficult to solubilize P-450, however, because of its instability and easy conversion to P-420.

C. Measurement of absolute absorption spectra using particulate preparations

As mentioned above, P-450 was found to be an unusual hemoprotein judging from results obtained by the difference spectrum technique. It showed only a very small spectral change on reduction, almost no spectral change in the visible region on binding with carbon monoxide, and an atypical CO-shift of the Soret peak to longer wavelengths in contrast to usual hemoproteins. Furthermore, although the binding of the ferrous form with EIC resulted in the formation of usual α- and β-peaks in the visible region, it also resulted in an unusual splitting of the Soret peak into two peaks at 430 and 455 nm. It was thus very hard to draw possible absolute spectra on the basis of the knowledge obtained.

The next effort was directed towards utilization of particulate preparations for aquisition of information on the absolute absorption spectra. Horie, Kinoshita and their co-workers[6,7] worked out three methods using adrenocortical mitochondria and rabbit liver microsomes. The occurrence of P-450 in adrenocortical mitochondria had already been described in 1964 by Harding and his co-workers[8] and this finding was traced and confirmed by Kinoshita *et al.*[9] The first of the three methods[6]) was to obtain the absolute spectra of sonicated mitochondrial particles by using, as the optical reference, decolorized particles which had been prepared by treatment with alkali followed by treatment with hydrogen peroxide. The opaque glass method of Shibata[10]) was employed to improve the optical balance. The resulting spectra possessed a considerable part of the spectral characteristics of P-450 because the sonicated particles did not contain cytochrome b_5, contained only relatively small amounts of cytochromes b, c_1, c, and $a+a_3$, and P-450 could account for more than 80% of the total protoheme. The second method[6]) was based on the fact that P-450 could be converted to P-420 by treatment with p-chloromercuribenzoate (PCMB)[11]) and that the resulting P-420 was unstable under aerobic conditions.[3-5]) The absorption spectra of P-450 were recorded by using, as the optical reference, particles whose P-450 had been destroyed by aerobic treatment with PCMB at 5°C. On the other hand, it was already known that the P-450 in liver microsomes could be inductively synthesized by administration of drugs to animals, whereas the amount of b_5 did not show any appreciable change during the induction, as found by Conney et al.[12]) and by other authors.[13,14]) Based on these observations, Kinoshita and Horie[7]) devised a third method. They obtained approximate spectra for P-450 by placing hepatic microsomes from phenobarbital-induced rabbits in the sample cuvette and those from untreated rabbits in the reference cuvette. In order to cancel out the contribution of cytochrome b_5 in the spectrum, the concentration of microsomes was adjusted so that both cuvettes contained equal concentrations of b_5.

This kind of technique was further improved and extended by other authors. Thus, Nishibayashi and Sato[15,16]) measured the absolute spectrum of P-450 in microsomal particles obtained from phenobarbital-induced rabbit liver microsomes by removing b_5 selectively. The selective removal was carried out in 25% glycerol by digestion with bacterial protease. The same preparation was decolorized with hydrogen peroxide and used as the optical reference (Fig. 3.11).

In a similar way, Hildebrandt et al.[17]) measured the absolute spectrum of the oxidized form of P-450. They found that the oxidized form in phenobarbital-induced rabbit liver microsomes showed a low spin-type absorption spectrum, whereas that in methyl-cholanthrene-induced rabbit liver microsomes showed a high spin-type absorption spectrum (Fig. 3.12).

Although the accuracy of the absolute spectra obtained with particulate preparations was limited by difficulties in performing the spectrophotometry of turbid materials, these studies provided much information about the absolute spectra of P-450 before solubilized preparations

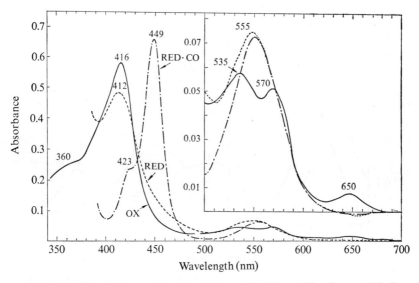

Fig. 3.11. Absolute absorption spectra of P-450 particles from rabbit liver microsomes.[15] P-450 particles containing 5.87 nmole P-450 per mg of protein were suspended in 0.1 M potassium phosphate buffer (pH 7.4) containing 30% glycerol (v/v). The turbidity of the sample was balanced with the bleached particles. The concentration of P-450 was 5.64 nmole per ml. ———, Oxidized P-450 particles; - - -, particles reduced with a small amount of solid $Na_2S_2O_4$. -·-·-, particles reduced with $Na_2S_2O_4$ and bubbled with CO for about 30 sec.

became available, as follows. (1) The reduced form has a broad single absorption peak at around 550 nm in the visible region. Because not only the change in shape of this peak is small but also the position of the maximum changes only slightly to a longer wavelength on binding with CO, no distinct peak appears in the visible region of the CO-difference spectrum. (2) As the shift of the Soret peak associated with oxidation-reduction is quite small compared with that in usual hemoproteins, only small peaks appear in the reduced minus oxidized difference spectrum. (3) The oxidized form of hepatic microsomal P-450 generally shows a ferric low spin-type spectrum. However, the reduced form shows a ferrous high spin-type spectrum like that of hemoglobin or peroxidases. (4) Dependent on the type of drugs administered to the animals, either ferric low spin- or ferric high spin-type P-450 can be inductively synthesized.

Another significance of the studies on the absolute spectra of particle-bound P-450 was that they offered a kind of standard spectrum which could be referred to in subsequent purification studies to examine possible alterations of P-450 during the solubilization and purification procedures.

78 MOLECULAR PROPERTIES

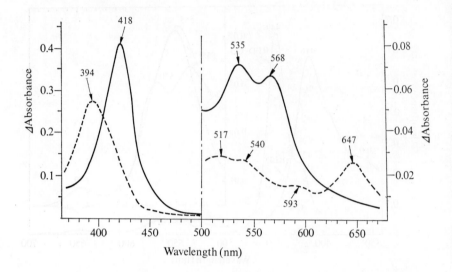

Fig. 3.12. Comparison of the absorption spectrum of the oxidized hemoprotein of liver microsomes induced by treatment of rabbits with 3-methylcholanthrene or phenobarbital. Microsomes from livers of control, PB- or 3-MC-treated rabbits were examined spectrophotometrically to determine the content of cytochrome b_5 using the procedure described by Omura and Sato.[4]. The spectrum of the additional pigment associated with microsomes from PB-treated rabbits (solid line) was determined by recording the spectral difference between a cuvette containing liver microsomes from PB-treated rabbits (2.2 mg protein per ml, 0.73 nmole cytochrome b_5 per mg, 2.92 nmole hemin per mg) minus a cuvette containing liver microsomes from a saline control rabbit (3.0 mg protein per ml, 0.51 nmole cytochrome b_5 per mg, 1.44 nmole hemin per mg). The additional pigment associated with microsomes from 3-MC treated rabbits (dashed line) was determined in a similar manner by recording the spectral difference between a cuvette containing liver microsomes from 3-MC treated rabbits (3.0 mg protein per ml, 0.79 nmole cytochrome b_5 per mg, 2.5 nmole hemin per mg) minus a cuvette containing liver microsomes from a corn oil control treated rabbit (3.0 mg protein per ml, 0.78 nmole cytochrome b_5 per mg, 1.6 nmole hemin per mg). Microsomes were diluted in 50 mM Tris buffer, pH 7.5, containing 15 mM KCl to the protein concentrations indicated. Temperature, 25°C. (Source: ref. 17. Reproduced by kind permission of Academic Press, Inc., U.S.A)

D. Spectral studies with solubilized preparations

The next step in the spectral study of P-450 was developed after its successful extraction and purification. Kinoshita et al.[18,19] in 1966, and subsequently Mitani and Horie,[20] extracted P-450 from sonicated mitochondrial particles obtained from bovine adrenal cortex, with phosphate buffer containing cholate, and partially purified this perparation by ammonium sulfate fractionation and gel filtration. Conversion to P-420 was effectively kept to a minimum by controlling the ratio of cholate to protein, and the small amounts of P-420 formed during the extraction procedure were destroyed by maintaining aerobic conditions in the presence of dithionite. The resulting preparation contained about 4 nmole protoheme per mg of protein. This stayed in the supernatant after centrifugation at 105,000 g for 90 min, and did not contain other cytochromes, non-heme iron, flavin, or labile sulfur. At liquid nitrogen temperature, the preparation showed ESR signals of low spin hemoprotein (microsomal Fe_x signals[21]). The preparation was practically free from contaminant P-420, and showed both steroid 11β-hydroxylation and cholesterol side-chain cleavage activity when tested in a reconstituted hydroxylation system.[22] The spectra of the preparation are shown in Fig. 3.13.

At about the same time, Nakamura and Otsuka,[23] and slightly later Schleyer et al.,[24] Isaka and Hall[25] and Jefcoate et al.,[26] also obtained solubilized preparations of adrenocortical mitochondrial P-450 and described their spectra. The solubilization of mitochondrial P-450 was thus achieved slightly earlier than that of microsomal P-450, due probably to the fact that mitochondrial P-450 is relatively more stable

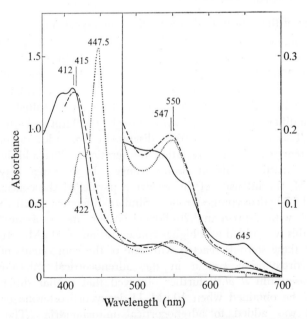

Fig. 3.13. Absorption spectra of a mixed spin-type P-450 preparation from adrenal cortex mitochondria.[20] The concentration of protein was approximately 2.7 mg/ml. The solvent was 0.1 M phosphate buffer (pH 7.0).——, Oxidized form; - - -, reduced with dithionite;···, carbon monoxide complex.

to cholate than microsomal P-450. The solubilization of microsomal P-450 was accomplished by the efforts of Coon and his associates,[27,28] Lu et al.,[29,30] Miyake et al.,[31] Mitani et al.,[32] Fujita et al.[33] and Imai and Sato.[34] Utilization of the stabilizing effect of glycerol on P-450 in the solubilization procedure played an important role in these studies.

P-450 was also obtained from sources other than animal tissues. Appleby[35] successfully extracted P-450 from *Rhizobium japonicum* (nitrogen-fixing bacteria of root nodules), Yoshida and Kumaoka[36] from yeast, and Katagiri et al.[37] from *Pseudomonas putida* grown in media containing camphor as sole carbon source. Among these, the P-450 from *Ps. putida* (P-450$_{CAM}$) has been highly purified and crystallized.[38]

As will be described later in this chapter, the spectral characteristics of solubilized P-450 preparations agree with those of membrane-bound P-450. Since absolute spectra became easily available after solubilization was attained, precise studies on the shape of the spectra, the spectral changes on binding with ligands or substrates, and the reaction of P-450 with oxygen or carbon monoxide became possible. As a result, much progress ensued in studies on the reaction mechanism and change of state of P-450 during catalysis.

E. Discovery of the "substrate-induced difference spectrum"

Another important event in the development of spectral studies on P-450 was the discovery of the so-called substrate-induced difference spectrum. In 1963, Narasimhulu et al.[39,40] reported that the addition of 17-hydroxyprogesterone to adrenocortical microsomes which had been clarified with Triton N101 (a non-ionic detergent) resulted in the formation of a difference spectrum having a peak at 390 nm and a trough at 420 nm. This difference spectrum disappeared on addition of dithionite but reappeared after oxidation by aeration. The addition of NADPH also caused a transient disappearance. In the range between 0.5 and 10 μM, the intensity of the spectrum paralelled the concentration of added 17-hydroxyprogesterone. Similar difference spectra were also observed with 5-pregnan-3,20-dione-17-ol or androst-4-ene-3,17-dione, when this was added at a higher concentration of 90 μM. It was assumed that these steroids reacted with one of the components of the steroid 21-hydroxylation system in the adrenocortical microsomes. Cooper, Narasimhulu et al.[41] further observed that similar difference spectra could be obtained when 10 μM 11-deoxycorticosterone or 11-deoxycortisol was added to adrenocortical mitochondria. The mitochondria, however, did not show any difference spectrum in the case of 17-hydroxyprogesterone addition.

In 1965 to 1966, in collaboration with Cooper and Narasimhulu,

Estabrook et al.[42,43] described that hepatic microsomes showed difference spectra on the addition of certain drugs. At about the same time, Imai and Sato[44] also reported substrate difference spectra for hepatic microsomes. When hexobarbital was added to the liver microsomes, the difference spectrum showed a peak at 385 nm and a trough at 420 nm (this was later termed a Type I difference spectrum). On the contrary, the addition of aniline caused a difference spectrum having a peak at 430 nm and a trough at 390 nm (this was later termed a Type II difference spectrum). When hepatic microsomal P-450 was inductively synthesized by the administration of phenobarbital to animals, the intensity of both the Type I and Type II difference spectra increased. However, when benzpyrene was used to stimulate the synthesis of P-450 instead of phenobarbital, the intensity of the aniline-difference spectrum increased but that of the phenobarbital-difference spectrum decreased. Although the pigment which combined with the substrate was uncertain, in the strict sense, it was believed to be P-450. This uncertainty originated from another possibility which could not be ruled out in those days, viz. that the substrate might not combine directly with P-450 but combine with another membrane component or with an assumed specificity-determining enzyme (so-called "true hydroxylase") and thus the substrate might affect the spectrum of P-450 via these structural factors or via certain macromolecular interactions. Later, however, when solubilized preparations became available, substrate difference spectra were also shown with such solubilized preparations.[20] Thus, the idea that substrates combine with P-450 itself obtained a sounder basis. In addition, the spectral studies of Mitani and Horie,[45] and also of Jefcoate and Gaylor,[46] revealed that substrate-induced difference spectra could be understood as high spin-low spin difference spectra of a ferric hemoprotein. This inference was also supported by ESR measurements.[45,47]

F. The spectral "abnormality" of P-450 and prospects for spectral studies

The history of research on the spectral properties of P-450 has been very eventful. It is thus interesting to examine the motive power which caused so many researchers to participate in the spectral study of P-450, and also to speculate on the future development of such studies. Moreover, it may be instructive to consider possible answers to these questions before actually entering into a precise description of the absorption spectra themselves.

P-450 apparently had unique and exceptional spectral characteristics compared with usual hemoproteins, and this uniqueness certainly drew the fascination of many researchers. With usual hemoproteins, there was a considerable regularity in the absorption spectra.[48,49] If a hemoprotein is an electron carrier (a cytochrome in its function), the oxidized form shows a ferric low spin-type (ferrihemochrome type) absorption spectrum and the reduced form shows a ferrous low spin-type (ferrohemochrome type) absorption spectrum. The change in valency of the heme iron is rapid but the reaction with added ligand does not occur at all or, at most, only very slowly. If a hemoprotein is an oxygen carrier (like hemoglobin) or an oxidizing enzyme (like peroxidase), the oxidized form shows a ferric high spin-type absorption spectrum and the reduced form shows a ferrous high spin-type absorption spectrum. Its velocity of ligand exchange is high. The spectral change on biniding with added ligands (e.g. the carbon monoxide-shift of the spectrum) follows a general rule.[50] Only P-450 appears to disobey these rules.

The abnormal spectral behavior of P-450 may be summarized as follows. (1) The spectral spin type of P-450 changes not only on binding with such ligands that cause a spin state change of usual hemoproteins (cyanide, azide, or fluoride) but also on binding with a large number of drugs, steroids and other substances (so-called substrates). Among the compounds which induce the substrate difference spectrum, it is not posible to distinguish any clear, common resemblance in molecular structure, and many of the compounds are quite unlikely to be a coordinant of heme iron. (2) The spectra undergo change not only on binding with substrate but also depending on the pH and in some cases, the ionic strength. (3) Both the ferric low spin form and the ferric high spin form change to ferrous high spin form on reduction, (4) When the reduced form reacts with carbon monoxide, the resulting spectral change in the visible region is unusually small, and α- and β-peaks of the absolute spectrum are not observable. It appears as if they are fused. In comparison with usual hemoproteins, the carbon monoxide-shift of the Soret band occurs towards longer wavelengths by as much as 30 nm, and the position of the Soret maximum of the CO-complex (450 nm) is at an almost unbelievably long wavelength for a protoheme-hemoprotein. (However, the last point may be understood in a different way. That is to say, the absorption bands of P-450 are located at longer wavelengths than in usual protoheme-hemoproteins due to the effect of its coordination structure, and the really unique feature is that the Soret band of the reduced form is located at unusually short wavelengths compared with those of other forms.[51] (5) On binding of EIC to the reduced form, a fission of the Soret band into two peaks at 430 and 455 nm appears, and the relative intensity of the two peaks varies with the pH and ionic strength.

Some of these unusual characteristics strongly suggest that the protein structure of P-450, especially the structure surrounding the heme, is not rigid but changes easily and dynamically dependent on the conditions. (For example, a change in the distance from the heme iron to the natural 6th coordinant,[52] an exchange of coordinants, or a change in electron density of the coordinants might occur by reduction or a change in pH.) Such easily variable structure is probably related to the catalytic function of this hemoprotein enzyme. One of the main efforts in the field of spectral studies will thus be directed towards elucidation of the 5th and 6th coordinants and another will be directed towards understanding the relation between spectral changes and changes in the structure surrounding the heme. The information obtained from studies of this "abnormal" hemoprotein will also need to be referred back to the case of other hemoproteins, and can thus be expected to prove valuable in deepening our understanding of hemoproteins in general.

Two topics have already been discussed by many authors: the coordination of $-S^-$ (thiolate anion, or mercaptide anion) to heme iron as a natural ligand, and the hydrophobic milieu of the heme in the protein molecule. The idea that the spectral characteristics of P-450 may be explained by assuming a sulfhydryl group as a natural coordinant was proposed by Murakami and Mason[53] as early as 1967. Recently, analytical results indicating that both cysteine and histidine residues are present close to heme in the P-450 molecule[54] have been reported. Model studies on this line have also been carried out.[47,51,55-59] Although the above idea has not yet been unequivocally established, many authors assume $-S^-$ as the 5th coordinant. On the other hand, a hydrophobic part is assumed to be present at a juxtaheme position on the basis of the fact that many of the substrates are hydrophobic compounds or compounds having a hydrophobic group in their molecule, and also that organic solvents at low concentrations affect the spectra of P-450.[60]

3.2.2 Difference Spectra

A. Common precautions necessary for measurement of the difference spectra

For accurate determinations of the difference spectra of P-450, many precautions are necessary. First, there should be no imbalance in conditions between the sample and the optical reference, except for the condition to be studied. Precise restrictions are as follows. (1) All imbalances of concentration, pH, ionic strength, etc., which may result from the addition of reagents, must be avoided. It should also be kept in mind that the addition of sparingly soluble substrate may cause an imbalance of transparency. (2) The effects of the organic solvents used for dissolving the substrate must be taken into consideration. Usually, the same amount of solvent is added to the reference; otherwise the obtained spectrum may be a solvent difference spectrum. Even at low concentrations, organic solvents often affect the intensity of substrate difference spectra.[45] (3) A pair of cuvettes having exactly the same length of light path should be used. Even with a very small difference in light path, a concentration difference spectrum may be obtained instead of the base line, and the recording of the desired difference spectrum will become difficult. This kind of trouble is often seen when a high concentration of the preparation is used. (4) When particulate preparations are measured, the particles may sediment in the cuvettes during determination. To prevent such sedimentation, either sonication or adjustment of the density of the medium is often effective. (5) When the concentration of the particulate preparation is high, the sensitivity of the spectrophotometer must be increased. It is better to employ a sensitive instrument specially designed for the measurement of turbid samples. Otherwise, in the range of wavelengths where the absorption of the preparation is high, recordings of simply electromechanical origin (one might say, ghost spectra) may be obtained. (6) In some preparations, other substances may be present which interfere with the recording of the difference spectrum by responding to added reagents. In the case of the CO-difference spectrum, other CO-binding hemoproteins interfere with the determination of P-450. Cytochrome oxidase, hemoglobin, and P-420 represent the practically important interfering substances. Cytochrome oxidase shows a large trough at 445 nm and a sharp peak at 430 nm in its CO-difference spectrum. If a large amount of cytochrome oxidase is present in the preparation, the peak of P-450 will therefore be hidden in the trough and will be difficult to detect. Hemoglobin shows a deep trough at around 430 nm and a sharp peak at 418 nm, rendering the determination of P-450 inaccurate. Similar interference can also be seen with P-420.

B. Carbon monoxide-difference spectrum

As already shown in Fig. 3.10, P-450 exhibits a large peak at about 450 nm, a deep trough at about 405 nm, and a broad, shallow trough (minimum at 480–520 nm) in the Soret region. In most instances a shoulder is seen at about 423 nm which is conspicuous when the preparation is contaminated with hemoglobin and/or P-420. Heavy contamination with hemoglobin or P-420 may even cause a distinct peak at 420 nm. The shoulder at 423 nm is almost absent or, if present, very small in carefully obtained particulate preparations or highly purified soluble preparations. Contamination by reduced hemoglobin or oxyhemoglobin can be detected by recording the CO-difference spectrum without the addition of dithionite, since they readily combine with carbon monoxide in the absence of a reducing agent. However, this technique is not applicable to methemoglobin or P-420. The Soret absorption maximum of P-450 varies slightly depending on the preparation used. For example, hepatic microsomal P-450 from untreated animals and that from phenobarbital-treated animals exhibit the maximum at 450 nm, whereas for hepatic microsomal P-450 from methylcholanthrene-treated animals the maximum is at 448 nm. These two kinds of P-450 were first described by Sladek and Mannering.[61] To distinguish them, the former is often called P-450 and the latter P-448 (or P_1-450). A difference in their respective catalytic activities has also been described.[62] With adrenocortical mitochondrial P-450, the Soret maximum is usually seen at around 448 nm. An example of the CO-difference spectrum of solubilized adrenocortical mitochondrial P-450[63] is given in Fig. 3.14.

In the visible region, two minor peaks are usually seen, one at about

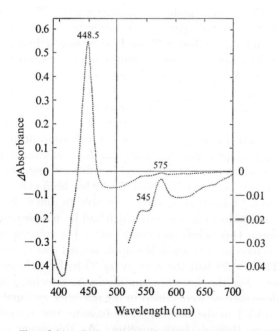

Fig. 3.14. Carbon monoxide-difference spectrum of a high spin-type P-450 preparation from adrenal cortex mitochondria.[63] Concentration of P-450, 7.27 μM; protein, 1.78 mg/ml. Solvent, 0.1 M phosphate buffer, pH 7.0. The lower curve in the visible region is drawn on a 10-fold expanded scale.

570 nm and the other at about 540 nm, and their intensities are extremely small. Slightly varied shapes in the visible region of the spectrum can be distinguished from the literature. In some preparations, contamination by P-420 or some other hemoprotein may have affected the spectral shape.

Either enzymatic (anaerobic) or chemical (with dithionite) reduction can be employed to form the carbon monoxide complex, although the intensity of the 450 nm peak is often smaller after enzymatic reduction than after the reduction by dithionite. In the determination of P-450 in microsomes, if dithionite is added prior to the bubbling of carbon monoxide, P-450 is often converted partially to P-420 by the peroxidation of microsomal lipids. Therefore, the carbon monoxide should be bubbled through prior to the reduction by dithionite. For more accurate determinations, one must replace the air in the anaerobic cuvette with carbon monoxide and tip the dithionite anaerobically. As the intensity of the 450 nm peak increases with time (up to 30 or 60 min) after the addition of dithionite, recordings for the purpose of quantitative determination must be repeated until a maximum is attained. The reaction of the reduced form with carbon monoxide is very rapid. The cause of the slow formation of the 450 nm peak is the slow reduction of P-450. It must be kept in mind that, when a Type II substrate is bound to P-450 (or the P-450 is in a low spin state), a longer time is required for complete reduction.[64,65] When the hemoglobin contamination is not serious, it is possible to calulate the hemoglobin concentration from the height of the 420 nm shoulder, and the concentration of P-450 can then be suitably corrected for the contamination,[66] as shown in Fig.3.15.

As stated above, when the preparations contain high concentrations of other CO-binding hemoproteins, the detection (or determination) of P-450 is usually difficult unless extraction and purification procedures are carried out. However, in some cases the difficulty has been overcome by other suitable means. Ghazarian *et al.*[67] were able to detect P-450 in mitochondria from the kidneys of chickens which had been maintained for 4 weeks on a vitamin D-deficient diet containing 1.1% calcium and then maintained for 6 days on a vitamin D_3-supplemented diet containing 3% calcium. This P-450 had the activity of 25-hydroxycholecalciferol-1α-hydroxylase. For the detection, antimycin A, ascorbate, and tetramethyl-*p*-phenylenediamine were added to the reference and malate and ascorbate were added to the sample. The baseline was recorded after bubbling nitrogen through both cuvettes. At this stage only cytochrome oxidase was reduced in the reference cuvette and both cytochrome oxidase and P-450 were reduced in the sample cuvette. Carbon monoxide was then bubbled through both cuvettes, and this resulted

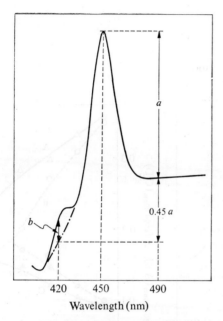

Fig. 3.15. Carbon monoxide-difference spectrum of dithionite-treated liver microsomes contaminated by hemoglobin.[66] The main peak at 450 nm is caused by the CO-compound of reduced P-450, whereas the shoulder around 420 nm is due to CO-hemoglobin. The absorbance increment between 450 and 490 nm ($\Delta A_{450-490\text{nm}}$), denoted as a, is used for calculation of the P-450 content ($\Delta\varepsilon$, 91 cm$^{-1}\cdot$mM^{-1}). Since $\Delta A_{490-420\text{nm}}$ was determined to be 0.45a in the CO-difference spectrum of a microsomal preparation essentially free from hemoglobin contamination, the height of the shoulder around 420 nm, denoted as b, can be estimated on the spectrum as illustrated. The hemoglobin content can be calculated from b, using a $\Delta\varepsilon$ value of 93 cm$^{-1}\cdot$mM^{-1}.

in the formation of a difference spectrum having a maximum at 450 nm. Such P-450 has been solubilized, examined, and found to be different from microsomal P-450.

C. Isocyanide-difference spectrum

Ethyl isocyanide (EIC)-difference spectra of bovine adrenal cortex mitochondria[9] are illustrated in Fig. 3.16. The EIC-difference spectrum of the reduced form of P-450 (solid line) shows a distinct α-peak at about 560 nm and a β-peak at about 530 nm in the visible region. These peaks indicate that the reduced form has a hemochrome type spectrum. In the Soret region, two are observed at 455 and 430 nm, which are thought to originate from two different states of EIC-ferrous P-450 complex.[68] The relative intensities of the 455 peak and the 430 peak vary with the conditions employed, especially with the pH and

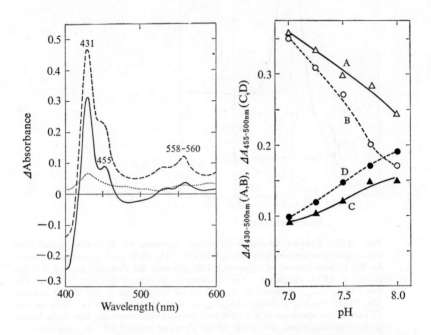

Fig. 3.16 (left). Ethyl isocyanide-difference spectra of sonicated mitochondrial fragments from bovine adrenal cortex.[9] Concentration of total nitrogen in the preparation, 0.3 mg/ml; final concentration of ethyl isocyanide (EIC) added, 10^{-3} M; final pH of the preparation mixed with EIC, 7.4.······, Oxidized preparation mixed with EIC minus oxidized;———, reduced with dithionite and mixed with EIC minus reduced;- - -, reduced with dithionite and mixed with EIC minus oxidized.

Fig. 3.17(right). Effect of pH on the ethyl isocyanide-difference spectra of reduced P-450 at low and high ionic strengths.[69] The difference spectra induced by 1.5 mM ethyl isocyanide in dithionite-treated liver microsomes (1.7 mg protein per ml) suspended in 0.05 M and 0.2 M potassium phosphate buffer were measured at various pH's. The intensities of the two peaks in the difference spectra, i.e. $\Delta A_{455-500\,nm}$ and $\Delta A_{430-500\,nm}$, were plotted against pH. Curve A, 430 nm peak in 0.05 M buffer; curve B, 430 nm peak in 0.2 M buffer; curve C, 455 nm peak in 0.05 M buffer; curve D, 455 nm peak in 0.2 M buffer.

ionic strength[68,69] (Fig. 3.17). The concentration of EIC is usually of the order of 10^{-3} M.

Similar difference spectra for the reduced form are also seen with pyridine and imidazole, but not with cyanide.[70] When alkyl isocyanide compounds of various carbon chain length are used, the relative intensities of the 455 peak and the 430 peak were found to vary depending on the carbon chain length.[71]

The EIC-difference spectrum of the oxidized form has peaks in both the visible and Soret regions, but their intensities are small.

D. Reduced minus oxidized difference spectrum

With purified preparations the reduced minus oxidized spectrum can be obtained simply by recording the difference, dithionite-added minus untreated preparation. With microsomes or submitochondrial particles, however, a slightly more complicated procedure is required. In the case of liver microsomes, the difference spectrum can be obtained by taking advantage of the fact that cytochrome b_5 is reduced aerobically with NADH whereas P-450 can be reduced by either dithionite or the anaerobic addition of NADPH. In usual practice, the [(dithionite-added) minus (NADH-added, aerobic)] or [(NADPH-added, anaerobic) minus (NADH-added, aerobic)] difference is recorded (Fig. 3.18).[66] As both the peaks in the Soret and visible regions are weak and broad,

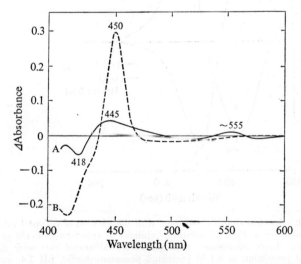

Fig. 3.18. Difference spectra of liver microsomes prepared from a phenobarbital-treated rabbit.[66] The microsomal concentration was 1.0 mg protein (3.51 nmole P-450) per ml. Curve A, dithionite minus NADH difference spectrum; curve B, CO-difference spectrum of the dithionite-treated sample.

the spectrum has no practical importance except for special cases. The peaks are usually observed at around 445 nm and 555 nm with liver microsomes. It should be noted that the shape of the reduced minus oxidized difference spectrum may vary depending on the spin state of the oxidized form.

E. Substrate-induced difference spectrum

Substrate-induced difference spectra in the Soret region are usually classified into three types, as proposed by Schenkman et al.,[72] on the basis of results obtained with liver microsomes. These are Type I, Type II, and modified Type II difference spectra. As shown in Fig. 3.19,[73] Type I has a maximum at 385–390 nm and a minimum at about 420 nm (dashed line). Almost symmetrical to Type I, Type II has a maximum at about 430 nm and a minimum at 390 nm (solid line). Slightly aside from the Type II difference spectrum, modified Type II spectra

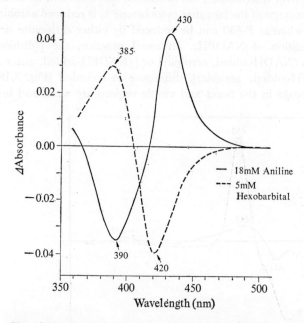

Fig. 3.19. Spectral changes associated with the addition of a Type I substrate (hexobarbital) or a Type II substrate (aniline) to cytochrome P-450 of liver microsomes. Liver microsomes from phenobarbital-treated rats were diluted to 5 mg of protein/ml in 0.1 M potassium phosphate buffer, pH 7.4, and the suspension was divided equally into two cuvettes. After establishment of a base-line of equal light absorbance, aniline (18 mM) or hexobarbital (5 mM) was added to the sample cuvette and the resulting difference spectrum was recorded. (Source: ref. 73 Reproduced by kind permission of University Park Press, U.S.A.)

show their maxima at 409–445 nm and minima at 365–410 nm.

The following substrates have been reported to cause the Type I difference: aminopyrine, phenobarbital, amobarbital, hexobarbital, SKF 525A (β-diethylaminoethyl diphenylpropylacetate, an inhibitor for the hydroxylation of drugs), DDT (dichlorodiphenyltrichloroethane), chloropromazine, N,N-dimethylaniline, testosterone, imipramine, and β-estradiol. As substrates causing the Type II difference, the following have been reported: nicotine, aniline, nicotinamide, DPEA (2,4-dichloro-6-phenyl phenoxyethylamine hydrochloride, an inhibitor of microsomal oxidation reactions; also called Lilly 32,391), pyridine, and p-aminophenol. As substrates causing the modified Type II difference, cortisol, corticosterone, acetoanilide, ethyl isocyanide, cyanide, and rotenone have been described.[72] The intensity of the difference spectra which appeared at the saturation concentrations varied from substrate to substrate, but it was proportional to the substrate concentration.

Similarly to hepatic microsomes, adrenocortical mitochondria, adrenocortical microsomes, and P-450$_{CAM}$ of *Pseudomonas putida* (active for the hydroxylation of camphor) also show substrate difference spectra.

TABLE 3.8. Substrate-binding specificity of P-450[20] [†1]

Source	\multicolumn{6}{c}{Bovine adrenal cortex}	Rat liver
Preparation	\multicolumn{2}{c}{P-450 preparation from mitochondria[†2]}	\multicolumn{2}{c}{Mitochondria}	\multicolumn{2}{c}{Microsomes}	Microsomes
Sub-\Concen- strate \tration	6.6 μM	63 μM	6.6 μM	63 μM	6.6 μM	63 μM	63 μM
Cholesterol	(I)	I	(−)	(−)	(−)	(−)	
20α-OH cholesterol	II	II			(−)	II	
Pregnenolone	II	II	II	II	(II)	II	I
17-OH pregnenolone	(−)	II	(−)	II	(−)	(II)	
Progesterone	I	I	(I)	I	I	I	I
Deoxycorticosterone	I	I	I	I	I	I	II
Deoxycortisol	I	I	I	I	(−)	I	
Corticosterone	(−)	(I)	(−)	(−)	(−)	(−)	II
Cortisol	(−)		(−)	(−)	(−)	(II)	
Cortisone	(−)	(−)	(−)	(−)	(−)	(−)	
Estradiol	(−)	II	II			II	I
Testosterone	I	I	I	I	(−)		
5α-Androstane-3,17-dione	I	I	I	I	I	I	I
Phenobarbital	2.5 mM II		2.5 mM II		2.5 mM II		I
Nicotinamide	5 mM (II)		5 mM II		5 mM (II)		II

[†1] I, Type I difference spectrum; II, Type II difference spectrum.
[†2] Mixed spin-type preparation.

The Type II difference spectra which appear on binding of adrenocortical mitochondial P-450 with steroids show the peak at 420 nm, and are rather similar to the modified Type II in the classification of hepatic microsomal difference spectra (Table 3.8). Most compounds which induce substrate difference spectra are substrates for the P-450-mediated monooxygenase reaction, but others (for example, inhibitors) are not. In a broader sense, however, those difference spectra which are induced by compounds other than substrates of the P-450-catalyzed reaction are also referred to as substrate-induced difference spectra.

Factors determining the type of difference spectrum are probably as follows: (1) the kind of substrate used, (2) the kind of P-450 or the source of the preparation, and (3) the state of the P-450 or the kind of pretreatment for the preparation. For example, β-estradiol induces a Type I difference spectrum with liver microsomes but induces a Type II spectrum with adrenocortical mitochondria (Table 3.8). A mixed spin-type P-450 preparation from adrenocortical mitochondria (preparation A in Table 3.9, having strong 11β-hydroxylation activity and some cholesterol side-chain cleavage activity) shows either Type I or Type II difference spectra depending on the steroids added. However, high spin type P-450 preparations (preparation B in Table 3.9) (having strong 20α-hydroxycholesterol side-chain cleavage activity, and assumed to be a complex with cholesterol) show only Type II difference spectra.[74]

TABLE 3.9. Substrate-binding specificity of P-450 preparations A and B from adrenal cortex mitochondria[†1]

Preparation	A [†2]		B [†3]	
Substrate\Concentration	6.6 μM	63 μM	6.6 μM	63 μM
Cholesterol	(I)	I	(−)	(−)
20α-OH cholesterol	II	II	II	II
Pregnenolone	II	II	II	II
17-OH pregnenolone	(−)	II	II	II
Progesterone	I	I	II	II
Deoxycorticosterone	I	I	(II)	II
Deoxycortisol	I	I	(II)	II
Corticosterone	(−)	(I)		
Cortisol	(−)			
Cortisone	(−)	(−)		
Estradiol	(−)	II		
Testosterone	I	I	(II)	II
5α-Androstane-3,17-dione	I	I	(II)	II

[†1] I, Type I difference spectrum; II, Type II difference spectrum.
[†2] Preparation A = mixed spin-type preparation.
[†3] Preparation B = high spin-type preparation.
(Source: ref. 74. Reproduced by kind permission of the New York Academy of Sciences, U. S. A.)

The intensity of the substrate difference spectrum is determined not only by the concentration of the substrate but is also affected by other coexistent substrates,[20] by the presence of an organic solvent,[45] by the pH,[45] by the ionic strength,[36] and by the presence of a detergent. The development of Type I difference spectra is usually rapid but that of Type II difference spectra is much slower. In the latter case, therefore, recordings often have to be carried out repeatedly until maximum development is attained. The intensity is often expressed by the peak to trough value, i.e. the algebraic difference between the optical density at the maximum and that at the minimum in the difference spectrum. Substrate difference spectra can be utilized as a convenient means of studying the substrate-binding specificity of P-450 preparations. In such studies, the competition between the substrate to be examined and a substrate which is already known to combine with the P-450 preparation is often measured. Preparations which are denatured and irreversibly converted to P-420 do not show any substrate difference spectrum. When substrate difference spectra are observed, one can assume that the P-450 preparation is catalytically active.

On binding of P-450 with a substrate, a difference spectrum appears not only in the Soret region but also in the visible region.[72,75] In both regions, the pattern of the substrate difference spectra agrees with that of the high spin-low spin difference spectra of common hemoproteins.[20,76] In other words, a Type I substrate difference spectrum corresponds to a change in the spin state of P-450 from low to high and a Type II (or a modified Type II) spectrum corresponds to a change from high to low.[45]

Another interesting problem concerns the kinds of chemical structure of the substrate which can induce each type of spectral change. Tentatively setting aside factors on the side of P-450, Yoshida and Kumaoka[77] proposed the following hypothesis for the interaction between substrate and P-450. That is, Type I difference spectra appear when a relatively large hydrocarbon residue of the substrate combines with the protein part of 419 nm type (low spin type) P-450. Type II difference spectra appear when an amino group of the substrate combines with heme iron of either 419 nm type (low spin type) or 394 nm type (high spin type) P-450. Modified Type II difference spectra represent a change which occurs when a hydroxyl group combines with the heme iron of 394 nm type (high spin type) P-450 resulting in conversion to the 416 nm type. Although this hypothesis requires further checking in various cases, it certainly represents a valuable step towards gaining a clearer understanding of the apparently complicated phenomena relating to substrate-induced spectral changes.

3.2.3 Absolute Absorption Spectra

The absolute absorption spectra of P-450 reported hitherto by various authors are identical in the macroscopic sense, but show differences when precise comparisons are made. These differences probably originate from differences in the source, kind and state of the P-450, and the purity of the preparation. In this section, absolute spectra of various solubilized preparations will be described.

A. Absolute absorption spectra of hepatic microsomal P-450

The spectra of the oxidized, reduced, and carbon monoxide-bound forms of P-450 purified by Imai and Sato[34] from phenobarbital-induced rabbit liver microsomes are shown in Fig. 3.20. The oxidized form has an α-peak at 570–580 nm and a β-peak at about 540 nm in the visible region, plus a Soret peak at 416 nm. A δ-peak is also seen at about 360 nm. The spectrum of the oxidized form is that of typical ferric low spin protoheme-hemoprotein (ferrihemochrome type). The dithionite-reduced form has a single peak at about 550 nm in the visible region, and a Soret peak at 414 nm. It should be noted that the intensity of this reduced Soret peak is smaller than that in the oxidized form, and

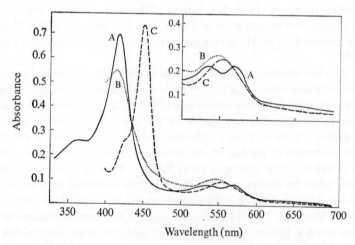

Fig. 3.20. Absolute absorption spectra of purified cytochrome P-450 from phenobarbital-induced rabbit liver microsomes.[34] A preparation containing 9.18 nmole cytochrome P-450 per mg protein was dissolved in 0.1 M potassium phosphate buffer (pH 7.25) containing 20% glycerol. The concentration of cytochrome P-450 was 6.46 μM (16.2 μM for the insert). Curve A, oxidized form; curve B, dithionite-reduced form; curve C, CO-compound of reduced form.

the position of the maximum is at a shorter wavelength (414 nm) than that of the oxidized Soret maximum (416 nm). Such spectral changes cannot take place unless a transition from low to high spin is associated with a change in the valency of iron. The spectral pattern of the reduced form is that of ferrous high spin hemoproteins, except that the Soret maximum is at a considerably shorter wavelength than that of usual high spin hemoproteins. The carbon monoxide-bound form shows a spectrum in the visible region which is almost identical to that of the reduced form, but its maximum is at a slightly longer wavelength than that of the reduced form. The maximum of the Soret peak is at 450 nm, and a very small shoulder is seen at 423 nm. This 423 nm shoulder occurs more or less in many preparations. However, since it shows a tendency to decrease with purification, and only a trace of the shoulder is observed with crystalline P-450$_{CAM}$, some authors have concluded that the shoulder is not innate.

The protein absorption to Soret absorption ratio (A_{280}/A_{416}) of a more highly purified preparation reported by Imai and Sato[78] was found to be 0.45. This preparation contained about 60,000 g protein per mole of heme, in other words 16–17 nmole heme per mg of protein. The spectra shown in Fig. 3.20 agree essentially with the spectra of particulate preparations from hepatic microsomes[15] (Fig. 3.11), and the purified preparation showed catalytic activity. It is therefore unlikely that any alteration in the structure of P-450 occurred during the purification procedure.

B. Absolute spectra of crystalline P-450$_{CAM}$ from *Pseudomonas putida*

According to the report of Yu *et al.*,[38] the oxidized form of P-450$_{CAM}$ shows a ferric low spin-type spectrum having an α-peak at 571 nm, β-peak at 538 nm, Soret peak at 417 nm, and a shoulder (δ-peak) at 365 nm. The absorption ratio of the protein peak to Soret peak was about 0.7 and the amount of protein per mole of heme was 46,000 g, corresponding to a heme content of about 20 nmole per mg protein. The dithionite-reduced form showed a maximum at 540 nm in the visible region and a Soret maximum at 411 nm. The carbon monoxide-bound form showed a maximum at 550 nm and a Soret maximum at 446 nm. On binding with the substrate, camphor, the Soret maximum of the oxidized form shifted to 391 nm and two peaks in the visible region appeared at about 646 and 500 nm. The latter peaks represent charge-transfer bands (sometimes called high spin bands). Thus, the spectrum of the oxidized form changed to typical ferric high spin type on binding with camphor. The oxidized form showed ESR signals characteristic of ferric low spin hemoprotein at liquid nitrogen or helium temperatures, and its camphor complex exhibited ESR signals characteristic of ferric high spin hemoproteins at liquid helium temperature.[47]

C. Absolute absorption spectra of adrenocortical mitochondrial P-450

Various such preparations have been reported. Generally speaking, those preparations which have strong steroid 11β-hydroxylase activity precipitate at relatively low concentrations of ammonium sulfate in the presence of cholate. They tend to form highly polymerized molecules, are soluble when a detergent is present, and show either low spin- or mixed spin-type absorption spectra in the oxidized state. Deoxycorticosterone causes Type I, and 20α-hydroxycholesterol causes Type II difference spectra in these P-450 preparations. On the other hand, preparations having strong cholesterol side-chain cleavage activity precipitate at relatively high concentrations of ammonium sulfate in the presence of cholate. They are soluble in the absence of a detergent, show mostly high spin-type absorption spectra in the oxidized state, and contain endogenously bound cholesterol.[79–81] Only Type II difference spectra are observed in these preparations. At present, the most purified preparations of adrenocortical mitochondrial P-450 contain about 80,000 g protein per mole of heme.[79–81]

(1) *Preparations showing mostly low spin type absorption spectra in the oxidized state*

A preparation has been described by Schleyer et al.[24] which was extracted and partially purified in the presence of a non-ionic detergent (Triton N-101). It contained 1.5–2 nmole protoheme per mg of protein, and its activity ratio of 11β-hydroxylation to cholesterol side-chain cleavage as measured in a reconstituted hydroxylation system was reported to be 2.5–5.0 : 1. The Soret maximum of the oxidized form was at 416 nm and that of the carbon monoxide-bound form was at 448 nm. The oxidized form did not react with fluoride or azide, and hardly reacted with cyanide. However, it could combine with deoxycorticosterone and other steroids and was transformed to a high spin type. Nitrogenous bases and alcohols caused Type II difference spectra.

(2) *Mixed spin type preparations*

A preparation has been reported by Mitani and Horie[20] which was extracted and partially purified in the presence of cholate. It contained 4 nmole protoheme per mg of protein, and showed activities for both 11β-hydroxylation and cholesterol side-chain cleavage.[22] The former activity was stronger than the latter. The preparation was optically clear in the absence of added detergent, but showed a tendency towards gradual turbidity on dilution or standing. Its absorption spectra are shown above in Fig. 3.13. The oxidized form had a Soret maximum

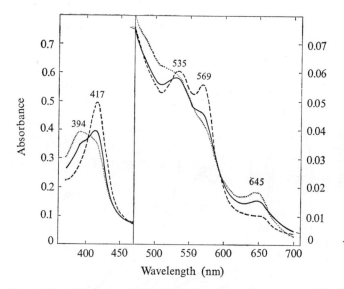

Fig. 3.21. Effect of steroid on the absolute absorption spectrum of the oxidized form of a mixed spin-type P-450 preparation from bovine adrenal cortex mitochondria.[45] ———, Control, without added steroid; ···, with deoxycorticosterone; - - -, with pregnenolone. To 3 ml of the preparation, 0.01 ml of 10 mM steroid in propylene glycol was added. The concentration of P-450 was approx. 3.1 μM.

at 412 nm and a distinct shoulder at 390 nm. The spectrum changed to a low spin type on the addition of pregnenolone or 20α-hydroxycholesterol, and the amount of high spin component increased on the addition of deoxycorticosterone[45] (Fig. 3.21; Tables 3.8 and 3.9). In parallel with the spectral change, changes in the ESR spectra were also observed.[45]

The preparation of Shikita and Hall[82,83] is known to have resembled Mitani-Horie's preparation in its spectral properties, but it was reported to contain 9–10 nmole heme per mg of protein. It showed higher activity for cholesterol side-chain cleavage than 11β-hydroxylation, the activity ratio of 11β-hydroxylation to cholesterol side-chain cleavage being 1:10–15.

(3) *Preparations showing mostly high spin-type absorption spectra in the oxidized state*

The first preparation showing a mostly high spin-type absorption spectrum was that reported by Jefcoate et al.[26] This preparation exhibited a high activity for cholesterol side-chain cleavage but its activity for 11β-hydroxylation was weak. Similar preparations were also described subsequently by Ando and Horie,[63] Isaka and Hall,[25]

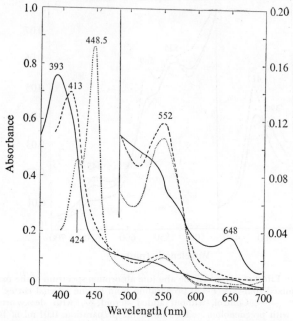

Fig. 3.22. Absorption spectra of a high spin-type P-450 preparation from bovine adrenal cortex mitochondria.[63] The concentration of P-450 was 7.27 μM; that o fthe protein was 1.78 mg/ml. The solvent was 0.1 M phosphate buffer (pH 7.0). ——, Oxidized form; - - -, reduced with dithionite; ···, carbon monoxide complex.

Takemori et al.[79] and Horie and Watanabe.[80,81] The absorption spectra of Ando and Horie's preparation are shown in Fig. 3.22. The oxidized form exhibited peaks at 393 and 648 nm and the spectrum was of mostly high spin type. By the addition of 20α-hydroxycholesterol, the spectrum changed almost completely to a low spin type, with a concomitant 3 to 4-fold increase in intensity of the low spin ESR signals. In a reconstituted hydroxylation system, this preparation strongly catalyzed side-chain cleavage of 20α-hydroxycholesterol, weakly catalyzed side-chain cleavage of cholesterol, but barely catalyzed 11β-hydroxylation of deoxycorticosterone. Horie and Watanabe's preparation, obtained by further purification of Ando-Horie's preparation, showed a protein to Soret absorption ratio of 1.2 in its oxidized state. It contained 80,000 g of protein per mole of heme, and 12.3 nmole heme per mg of protein. Its activity was, however, lower than that of Ando-Horie's preparation. The preparation reported by Takemori et al.[79] also contained 80,000 g protein per mole of heme. High spin-type preparations are known to exhibit only Type II difference spectra (Table 3.9).

D. Absolute absorption spectra of P-450 from corpus luteum mitochondria

The presence of P-450 in corpus luteum mitochondria was first described by Yohro and Horie,[84] and a soluble preparation has been reported by McIntosh et al.[85] Its spectral properties and activity are similar to those of high spin-type preparations from adrenocortical mitochondria.

E. Absolute spectra of *Rhizobium japonicum* P-450

By varying the cultural conditions, Appleby et al.[35,86] successfully obtained a series of P-450 preparations from *Rhizobium japonicum*, the spin states of the oxidized form of which were different. The spectra of the reduced forms and CO-bound forms of these preparations, however, did not show appreciable variations.

F. Absolute spectra of P-450 from yeast

The absolute spectra of a solubilized preparation obtained from yeast microsomes have been described by Yoshida and Kumaoka.[36] They were almost identical to the spectra of P-450 preparations obtained from hepatic microsomes.

G. Absolute absorption spectra of complexes of P-450 with ligands, low temperature spectra of P-450, and spectra of P-420

The absolute spectrum of reduced P-450 complex with EIC has been described by Miyake et al.[31] and also by Imai and Sato[78] The Soret peak exhibited fission into two peaks, at 455 nm and 430 nm, of which the former was found to be intensified by raising the pH value. The absolute spectrum in the visible region also changed with the pH. It was of ferrohemochrome type, although less typical than the difference spectrum. Concerning the splitting of the Soret band by EIC, a model study has been reported.[87] Miyake et al.[31] also described the absolute spectra of the nitrogen monoxide complex of the oxidized and reduced forms of P-450. Yoshida and Kumaoka[36] have reported the absolute spectrum of an aniline complex of the oxidized form of yeast P-450. Its maximum was at 421 nm, but in the aniline-difference spectrum the maximum was at 427 nm and a trough was present at 390 nm. Yu and Gunsalus et al.[38] have described absolute spectra of the oxidized form of *Ps. putida* P-450$_{CAM}$ complexes with 1-phenylimidazole and β-mercaptoethanol. In the case of 1-phenylimidazole, the Soret maximum shifted from 417 to 422 nm and the height of the Soret peak decreased slightly. In the case of β-mercaptoethanol, the Soret band showed fission into three peaks at 380, 417, and 460 nm. This fission did not occur in the presence of the substrate, camphor. The absolute spectrum of the oxygenated form of P-450 has been reported by Ishimura et al.[88] and by Gunsalus et al.[89]

Using a mixed spin-type preparation from adrenal cortex mitochondria, Mitani and Horie[45] have measured low temperature spectra of the deoxycorticosterone complex (high spin type) and of the pregnenolone complex (low spin type). By lowering the temperature from room to liquid nitrogen temperature, a sharpening of the absorption bands was observed but no temperature-dependent change in the spin state could be detected. On the other hand, Peisach and Appleby[90] have studied low temperature spectra of *Rhizobium* P-450c (essentially low spin type). According to their results, when phenobarbital was bound to this P-450 a slight increase in high spin component occurred at room temperature and this complex changed to an essentially low spin type at liquid nitrogen temperature. In the absence of phenobarbital, *Rhizobium* P-450 showed no evidence of thermal equilibrium between high and low spin forms.

Absolute spectra of P-420 had already been described in 1964 by Omura and Sato[4] using a preparation from liver microsomes. At pH 7, the oxidized form of P-420 had its Soret maximum at 414 nm and showed a ferrihemochrome-type spectrum. The reduced form showed a ferrohemochrome-type spectrum with an α-maximum at 559 nm, β-maximum at 530 nm, and Soret maximum at 427 nm, The carbon monoxide complex had its Soret maximum at 420 nm, and two rather broad peaks were present in the visible region. An examination of the pattern of spectral change from the reduced form to the CO-bound form revealed that the structure around the heme was that of denatured protein hemochromes.[50] P-420 shows no spectral "abnormality".

3.2.4 Extinction Coefficients

Molar extinction coefficients are of practical value since if the coefficient is known, P-450 can be determined by simple spectrophotometry. As described above, the spectra of P-450 often change under the influence of various conditions. Thus, the application of extinction coefficients for determination must be performed with care. Furthermore, it is not guaranteed that all kinds of P-450 have the same extinction coefficient. Nevertheless, the determination procedure using the CO-difference spectrum and molar extinction coefficient increment is reliable and has been conveniently used. Since P-450 is a hemoprotein, the molar extinction coefficient should be determined and described on the basis of the molar (or millimolar) concentration of heme. Coefficients described on the basis of protein concentration (either molar or wt/vol) are not usually accurate and are of limited use.

A. The difference in millimolar extinction coefficient increments in CO-difference spectra between 450 and 490 nm

In the case of liver microsomes, a value of 91 cm^{-1}·mM^{-1} has been described by Omura and Sato[3-5] for the difference in millimolar extinction coefficient increments between 450 and 490 nm in the CO-difference spectrum. This has been considered to be the most reliable value and is widely used. The corresponding value for a purified preparation obtained from hepatic microsomes has recently been determined as 92 cm^{-1}·mM^{-1}.[78] These two values thus agree very closely. The value, 91, is often applied also to microsomes from other organs (and sometimes even to mitochondrial preparations) when reliable values are not available. However, different values have been described by other authors. Using Triton N101 solubilization, Fujita et al.[33] prepared P-450 from phenobarbital-induced micosomes and P-448 from methylcholanthrene-induced microsomes, and described a $\Delta\varepsilon_{mM}^{450-490nm}$ value of 58 cm^{-1}·mM^{-1} for P-450 and a $\Delta\varepsilon_{mM}^{448-490nm}$ value of 78 cm^{-1}·mM^{-1} for P-448, the former value (58) being about two thirds the corresponding value obtained by Omura and Sato.

Concerning adrenocortical mitochodrial P-450, Mitani and Horie[20] have calculated a $\Delta\varepsilon_{mM}^{448.5-490nm}$ value of 100 cm^{-1}·mM^{-1} for the mixed spin-type preparation, and Ando and Horie[63] have described a value of 85 cm^{-1}·mM^{-1} for the high spin-type preparation. The value reported by Shikita and Hall[82] for their mixed spin-type preparation was 185 cm^{-1}·mM^{-1} on the basis of the concentration of polymer molecule (molecular weight 850,000). According to their data, 850,000 g protein contained about 8 mole of protoheme. Thus, the corresponding value on the basis of heme would be 23 cm^{-1}·mM^{-1}.

Concerning the P-450$_{CAM}$ of Pseudomonas putida, a calculation from the spectrum reported by Yu et al.[38] has indicated that the value of $\Delta\varepsilon_{mM}^{446-490nm}$ was probably about 77 cm^{-1}·mM^{-1}.

B. Absolute extinction coefficient of the Soret maximum of CO-complex of P-450

A value of 118 cm^{-1}·mM^{-1} (448.5 nm) has been described by Ando and Horie[63] for high spin-type preparations obtained from adrenocortical mitochondria, and value of 106 cm^{-1}·mM^{-1} (446 nm) has been described by Yu et al.[38] for crystalline P-450$_{CAM}$.

C. Other absolute extinction coefficients

Ando and Horie[63] have analyzed high spin-type preparations obtained from adrenal cortex mitochondria, and described a value of 103 cm$^{-1}\cdot$mM^{-1} (393 nm) for the Soret maximum of the ferric high spin form and a value of 121 cm$^{-1}\cdot$mM^{-1} (416 nm) for the Soret maximum of the 20α-hydroxycholesterol complex (ferric low spin form). Concerning the crystalline P-450$_{CAM}$ reported by Yu et al.,[38] the following values have been given for the Soret maxima: ferric low spin form = 105 cm$^{-1}\cdot$mM^{-1} (417 nm), ferric high spin form (camphor complex) = 91 cm$^{-1}\cdot$mM^{-1} (391 nm), and ferrous form = 71 cm$^{-1}\cdot$mM^{-1} (411 nm). As for the Triton N101 solubilized microsomal preparations reported by Fujita et al.[33], the following values have been given for the Soret maxima: ferric low spin form = 104 cm$^{-1}\cdot$mM^{-1} (418 nm, P-450), and ferric low spin form = 120 cm$^{-1}\cdot$mM^{-1} (419 nm, P-448). Fujita et al.[33] and Yu et al.[38] have described extinction coefficients for the maxima in the visible region.

Millimolar extinction coefficients of the Soret peak of ferric protoheme-hemoprotein enzymes are usually around 100 cm$^{-1}\cdot$mM^{-1}. Thus, in this respect, P-450 is "normal".

3.2.5 Photochemical Action Spectrum of P-450

The photochemical action spectrum method was successfully elaborated by Warburg and his co-workers[91] by taking advantage of the fact that carbon monoxide complexes of heme compounds are dissociated by irradiation with light, and was utilized to demonstrate the spectrum of cellular respiratory enzymes. The respiratory enzyme (cytochrome oxidase, a hemoprotein) is scarcely active in an atmosphere of suitable CO/O_2 ratio. However, on irradiation with light, the enzyme is released from inhibition by CO and becomes reactive to O_2. The ability of light to release the enzyme from CO-inhibition varies with the wavelength. At wavelengths where the carbon monoxide complex of the respiratory enzyme absorbs more light energy, the extent of release from inhibition is high, resulting in increased oxidase activity. Therefore, the curve obtained by plotting the activity against the wavelength of the irradiated light agrees with the absorption spectrum of the CO-complex of the respiratory enzyme.

In 1963, Estabrook et al.[92] reported that the photochemical action spectrum of steroid 21-hydroxylation catalyzed by adrenocortical microsomes showed its maximum at 450 nm (Fig. 3.23). Roughly 100 times more intense irradiation was necessary for the determination than in

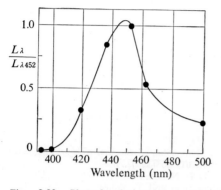

Fig. 3.23. Photochemical action spectrum for the light reversal of CO-inhibition of steroid hydroxylation. Each flask contained 6.2 mg of microsomal protein. Incubation time was 15 min, at 25°C. The rate of cortexolone formation in the absence of CO was 1.2 nmole/min/mg protein. The flasks for studying the CO-inhibition were gassed with a mixture of 4.4% O_2 and 9.5% CO in nitrogen (CO/O_2=2.2). The results are plotted in terms of relative light sensitivity $L\lambda/L\lambda_{452}$, vs, wavelength of the light applied. L was calculated from the equation

$$L = \frac{1}{i} \frac{K_h - K_d}{K_d} = \frac{1}{i} \frac{\Delta K}{K_d},$$

were i is the light intensity expressed as mole quanta.cm^{-2}.min^{-1} at the wavelength λ and K_h and K_d are the partition constants with and without illumination, respectively. K_d was 0.64.
(Source: ref. 92. Reproduced by kind permission of Springer-Verlag, W. Germany.)

the case of Warburg's respiratory enzyme. Using the powerful irradiation of liver microsomes obtained with a xenon lamp, Estabrook et al.[42,93–95] further studied the action spectra (or reversal of CO-inhibition) of microsomal hydroxylation of aniline, oxidative demethylation of codeine, hydroxylation of testosterone, and hydroxylation of acetanilide. They were able to show by these studies that P-450 is the terminal oxidase of these monooxygenase reactions. Subsequently, Cooper et al.[96,97] also proved that adrenocortical mitochondrial P-450 is the terminal oxidase of the steroid 11β-hydroxylation system using the action spectrum technique. Thus, by the photochemical action spectrum method (a functional test for hemoproteins), the role played by P-450 was elucidated before the purification of P-450 and reconstitution of the hydroxylase system were accomplished.

References

Section 3.2
1. M. Klingenberg, *Arch. Biochem. Biophys.*, **75**, 376 (1958).
2. D. Garfinkel *ibid.*, **77**, 493 (1958).
3. T. Omura and R. Sato, *J. Biol. Chem.*, **237**, PC1375 (1962).
4. T. Omura and R. Sato, *ibid.*, **239**, 2370 (1964).
5. T. Omura and R. Sato, *ibid.*, **239**, 2379 (1964).
6. S. Horie, T. Kinoshita and N. Shimazono, *J. Biochem. (Tokyo).*, **60**, 660 (1966).
7. T. Kinoshita and S. Horie, *ibid.*, **61**, 26 (1967).
8. B. W. Harding, S. H. Wong and D. H. Nelson, *Biochim. Biophys. Acta*, **92**, 415 (1964).
9. T. Kinoshita, S. Horie, N. Shimazono and T. Yohro, *J. Biochem. (Tokyo)*, **60**, 391 (1966).
10. K. Shibata, *Methods of Biochemical Analysis* (ed. D. Glick), vol. VII, p. 77, Interscience (1959).
11. T. Omura, S. Narasimhulu and R. W. Estabrook, *Oxidases and Related Redox Systems* (ed. T. E. King, H. S. Mason and M. Morrison), vol. 2, p. 876, John Wiley (1965).
12. A. H. Conney, C. Davison, R. Gastel and J. J. Burns, *J. Pharmacol. Exptl. Ther.*, **130**, 1 (1960).
13. H. Remmer and H. J. Merker, *Science*, **142**, 1657 (1963).
14. L. Ernster and S. Orrenius, *Fed. Proc.*, **24**, 1190 (1965).
15. H. Nishibayashi and R. Sato, *J. Biochem. (Tokyo)*, **63**, 766 (1968).
16. H. Nishibayashi, T. Omura, R. Sato and R. W. Estabrook, *Structure and Function of Cytochromes* (ed. K. Okunuki, M. D. Kamen and I. Sekuzu), p. 658, University of Tokyo Press (1968).
17. A. Hildebrandt, H. Remmer and R. W. Estabrook, *Biochem. Biophys. Res. Commun.*, **30**, 607 (1968).
18. T. Yohro, T. Kinoshita and S. Horie, *Seikagaku* (Japanese), **38**, 552 (1966).
19. T. Kinoshita, *Tokyo J. Med. Sci.* (Japanese), **75**, 202 (1967).
20. F. Mitani and S. Horie, *J. Biochem. (Tokyo)*, **65**, 269 (1969).
21. Y. Hashimoto, T. Yamano and H. S. Mason, *J. Biol. Chem.*, **237**, PC3843 (1962).
22. F. Mitani and S. Horie, *J. Biochem. (Tokyo)*, **68**, 529 (1970).
23. Y. Nakamura and H. Otsuka, 19th Symp. Enzyme Chemistry, Kanazawa (Japanese) (1968). Abstracts, p. 27
24. H. Schleyer, D. Y. Cooper and O. Rosenthal, *J. Biol. Chem.*, **247**, 6103 (1972).
25. S. Isaka and P. F. Hall, *Biochem. Biophys. Res. Commun.*, **43**, 747 (1971).
26. C. R. Jefcoate, R. Hume and G. S. Boyd, *FEBS Lett*, **9**, 41 (1970).
27. A. Y. H. Lu and M. J. Coon, *J. Biol. Chem.*, **243**, 1331 (1968).
28. A. Y. H. Lu, K. W. Junk and M. J. Coon, *ibid.*, **244**, 3714 (1969).
29. A. Y. H. Lu. and W. Levin, *Biochem. Biophy. Res. Commun.*, **46**, 1334 (1972).
30. W. Levin, A. Y. H. Lu, D. Ryan, S. West, R. Kuntzman and A. H. Conney, *Arch. Biochem. Biophys.*, **153**, 543 (1972).
31. Y. Miyake, J. L. Gaylor and H. S. Mason, *J. Biol. Chem.*, **243**, 5788 (1968).
32. F. Mitani, A. P. Alvares, S. Sassa and A. Kappas, *Mol. Pharmacol.*, **7**, 280 (1971).
33. T. Fujita, D. W. Shoeman and G. J. Mannering, *J. Biol. Chem.*, **248**, 2192 (1973).
34. Y. Imai and R. Sato, *J. Biochem. (Tokyo)*, **75**, 689 (1974).
35. C. A. Appleby, *Structure and Function of Cytochromes* (ed. K. Okunuki, M. D. Kamen and I. Sekuzu), p. 666, University of Tokyo Press (1968).
36. Y. Yoshida and H. Kumaoka, *J. Biochem. (Tokyo)*, **71** 915 (1972).
37. M. Katagiri, B. N. Ganguli and I. C. Gunsalus, *J. Biol. Chem.*, **243**, 3543 (1968).
38. C. A. Yu, I. C. Gunsalus, M. Katagiri, K. Suhara and S. Takemori, *ibid.*, **249**,

94 (1974).
39. S. Narasimhulu, *Fed. Proc.* **22**, 530 (1963).
40. S. Narasimhulu, D. Y. Cooper and O. Rosenthal, *Life Sci.*, **4**, 2101 (1965).
41. D. Y. Cooper, S. Narasimhulu, A. Slade, W. Raich, O. Foroff and O. Rosenthal, *ibid.*, **4**, 2109 (1965).
42. R. W. Estabrook, J. B. Schenkman, W. Cammer, H. Remmer, D. Y. Cooper, S. Narasimhulu and O. Rosenthal, *Biochemical and Chemical Aspects of Oxygenases* (ed. K. Bloch and O. Hayaishi), p. 153, Maruzen (1966).
43. H. Remmer, J. Schenkman, R. W. Estabrook, H. Sasame, J. Gillette, S. Narasimhulu, D. Y. Cooper and O. Rosenthal, *Mol. Pharmacol.*, **2**, 187 (1966).
44. Y. Imai and R. Sato, *Biochem. Biophys. Res. Commun.*, **22**, 620 (1966).
45. F. Mitani and S. Horie, *J. Biochem. (Tokyo)*, **66**, 139 (1969).
46. C. R. E. Jefcoate and J. L. Gaylor, *Biochemistry*, **8**, 3464 (1969).
47. R. Tsai, C. A. Yu, I. C. Gunsalus, J. Peisach, W. Blumberg, W. H. Orme-Johnson and H. Beinert, *Proc. Natl. Acad. Sci. U.S.A.*, **66**, 1157 (1970).
48. R. J. P. Williams, *The Enzymes* (ed. P. D. Boyer, H. Lardy and K. Myrbaeck), vol. 1, p. 391, Academic press (1959).
49. S. Horie, *J. Biochem. (Tokyo)*, **57**, 147 (1965).
50. S. Horie, *ibid.*, **56**, 113 (1964).
51. T. Watanabe and S. Horie, *ibid.*, **79**, 829 (1976).
52. Y. Ichikawa and T. Yamano, presented at the 39th Ann. Mtg. Japan. Biochem. Soc. *Seikagaku* (Japanese), **39**, 756 (1967).
53. K. Murakami and H. S. Mason, *J. Biol. Chem.*, **242**, 1102 (1967).
54. K. Dus, W. J. Litchfield, A. G. Miguel, T. A. van der Hoeven, D. A. Haugen, W. L. Dean and M. J. Coon, *Biochem. Biophy. Res. Commun.*, **60**, 15 (1974).
55. H. A. O. Hill, A. Roeder and R. J. P. Williams, *Naturwissenschaften*, **57**, 69 (1970).
56. J. O. Stern and J. Peisach, *J. Biol. Chem.* **249**, 7495 (1974).
57. S. Koch, S. C. Tang, R. H. Holm and R. B. Frankel, *J. Am. Chem. Soc.*, **97**, 914 (1975).
58. S. Koch, S. C. Tang, R. H. Holm, R. B. Frankel, and J. A. Ibers, *J. Am. Chem. Soc.*, **97**, 916 (1975).
59. J. P. Collman and T. N. Sorrell, *J. Am. Chem. Soc.*, **97**, 4133 (1975).
60. Y. Ichikawa, T. Uemura and T. Yamano, *Structure and Function of Cytochromes* (ed. K. Okunuki, M. D. Kamen and I. Sekuzu), p. 634, University of Tokyo Press (1968).
61. N. E. Sladek and G. J. Mannering, *Biochem. Biophys. Res. Commun.*, **24**, 668 (1966)
62. A. Y. H. Lu, R. Kuntzman, S. West, M. Jackson and A. H. Conney, *J. Biol. Chem.*, **247**, 1727 (1972).
63. N. Ando and S. Horie, *J. Biochem. (Tokyo)*, **72**, 583 (1972).
64. P. L. Gigon, T. E. Gram and J. R. Gillette, *Mol. Pharmacol.*, **5**, 109 (1969).
65. N. Ando and S. Horie, *J. Biochem. (Tokyo)*, **70**, 557 (1971).
66. H. Nishibayashi and R. Sato, *ibid.*, **61**, 491 (1967).
67. J. G. Ghazarian, C. R. Jefcoate, J. C. Kuntson, W. H. Orme-Johnson and H. F. DeLuca, *J. Biol. Chem.*, **249**, 3026 (1974).
68. Y. Imai and R. Sato, *J. Biochem. (Tokyo)*, **63**, 370 (1968).
69. Y. Imai and R. Sato, *ibid.*, **63**, 270 (1968).
70. Y. Imai and R. Sato, *ibid.*, **62**, 464 (1967).
71. Y. Ichikawa and T. Yamano, *Seikagaku* (Japanese), **39**, 756 (1967).
72. J. B. Schenkman, H. Remmer and R. W. Estabrook, *Mol. Pharmacol.*, **3**, 113 (1967).
73. J. A. Peterson, Y. Ishimura, J. Baron and R. W. Estabrook, *Oxidases and Related Redox Systems* (ed. T. E. King, H. S. Mason and M. Morrison), vol. 2, p. 565, University Park Press (1971).
74. F. Mitani, N. Ando and S. Horie, *Ann. N. Y. Acad. Sci.*, **212**, 208 (1973).
75. J. A. Whysner, J. Ramseyer and B. W. Harding, *J. Biol. Chem.*, **245**, 5441 (1970).
76. S. Horie and M. Morrison, *Oxidases and Related Redox Systems* (ed. T. E. King, H. S.

Mason and M. Morrison), vol. 2, p. 659, John Wiley (1964).
77. Y. Yoshida and H. Kumaoka, *J. Biochem. (Tokyo)*, **78**, 455 (1975).
78. Y. Imai and R. Sato, *Biochem. Biophy. Res. Commun.*, **60**, 8 (1974).
79. S. Takemori, K. Suhara, M. Hashimoto, Y. Ikeda and M. Katagiri, *Seikagaku* (Japanese), **46**, 369 (1974).
80. S. Horie and T. Watanabe, IVth Int. Congr. Hormonal Steroids, Mexico City (1974). Symp. Abstracts, S-10(2).
81. S. Horie and T. Watanabe, *J. Steroid Biochem.*, **6**, 401 (1975).
82. M. Shikita and P. F. Hall, *J. Biol. Chem.*, **248**, 5598 (1973).
83. M. Shikita and P. F. Hall, *ibid.*, **248**, 5605 (1973).
84. T. Yohro and S. Horie, *J. Biochem. (Tokyo)*, **61**, 515 (1967).
85. E. N. McIntosh, F. Mitani, V. I. Uzgiris, C. Alonso and H. A. Salhanick, *Ann. N.Y. Acad. Sci.*, **212**, 392 (1973).
86. C. A. Appleby and R. M. Daniel, *Oxidases and Related Redox Systems* (ed. T. E. King, H. S. Mason and M. Morrison), vol. 2, p.515, University Park Press (1973).
87. Y. Imai and R. Sato, *Structure and Function of Cytochromes* (ed. K. Okunuki, M. D. Kamen and I. Sekuzu), p. 626, University of Tokyo Press (1968).
88. Y. Ishimura V. Ullrich and J. A. Peterson, *Biochem. Biophys. Res. Commun.*, **42**, 140 (1971).
89. I. C. Gunsalus, C. A. Tyson and J. D. Lipcomb, *Oxidases and Related Redox Systems* (ed. T. E. King, H. S. Mason and M. Morrison), vol. 2, p. 583, University Park Press (1973).
90. J. Peisach and C. A. Appleby, *ibid.*, p. 486 (1973).
91. O. Warburg and E. Negelein, *Biochem. Z.* **214**, 64 (1929).
92. R. W. Estabrook, D. Y. Cooper and O. Rosenthal, *ibid.*, **338**, 741 (1963).
93. D. Y. Cooper, S. Narasimhulu, O. Rosenthal and R. W. Estabrook, *Oxidases and Related Redox Systems* (ed. T. E. King, H. S. Mason and M. Morrison), vol. 2, p. 856, John Wiley (1965).
94. D. Y. Cooper, S. Levin, S. Narasimhulu, O. Rosenthal R. W. Estabrook, *Science*, **147**, 400 (1965).
95. T. Omura, R. Sato, D. Y. Cooper, O. Rosenthal and R. W. Estabrook, *Fed. Proc.*, **24**, 1181 (1965).
96. D. Y. Cooper. B. Novack, O. Foroff, A. Slade, E. Saunders, S. Narasimhulu and O. Rosenthal, *ibid.*, **26**, 341 (1967).
97. D. Y. Cooper and O. Rosenthal, *ibid.*, **25**, 765 (1965).

3.3 Magnetic Properties

3.3.1 History and an Outline of Microsomal Fe_x

While the absorption at 450 nm of reduced cytochrome P-450 under carbon monoxide is unusual among known *b* type cytochromes, the electron paramagnetic resonance (EPR) absorption of this cytochrome is typical and easily detectable. The EPR signal is observable at the lowest concentration among physiologically functional cytochromes. However, before the EPR signal of P-450 was established, some confusion arose because the signal was first assigned to a new iron protein, microsomal Fe_x, designated by H.S. Mason. One of the authors (T.Y.) had joined Dr. Mason's laboratory in 1961, where the new EPR signal was being observed using rabbit liver microsomes. At about the same time, Omura and Sato published details of their discovery of P-450 in liver microsomes.[1]

One of the advantages of EPR observations is their ability to describe the signals of radicals or transition-metal enzymes even in the presence of many optically non-transparent substances such as proteins, lipids and pigments. Thus, the EPR signal of microsomal Fe_x was detected in homogenates of liver tissues, and more clearly in the microsomes, as illustrated in Fig. 3.24.[2]

A set of triplet g values was noticed: these were 2.41, 1.91 (described as s_m, the slope maximum, in the original paper), and 2.25. To ascribe the triplet g values to a hemoprotein, we had to have some suitable experimental technique and theoretical background. For the former, the anaerobic EPR tubes designed by Beinert et al.[3] were useful because

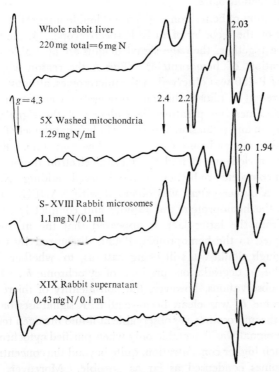

Fig. 3.24. EPR spectra of whole rabbit liver and homogenate fractions ($T=$ 113K). The samples were suspended in 0.2 ml of 0.01 M Tris-phosphate buffer, pH 8.2.
(Source: ref. 2. Reproduced by kind permission of J. Wiley & Sons, Inc., U.S.A.)

strictly anaerobic conditions were required, such as repeated evacuation and flushing with pure nitrogen. The triplet signals were diminished completely in parallel by careful dithionite titration. The temperature dependency of the triplet signals was very steep and was hardly observable over 150 K at a microsomal protein concentration of 20 mg/ml. When the logarithms of the signal heights were plotted against T, the temperature in Kelvin units, according to the method of Ehrenberg,[4] the α-values in $S=S_0 \exp(-T/\alpha)$, where S represents signal height, were the same for 49 K with respect to the three g values.[2] Theoretically, an EPR signal with triplet g values, such as a type centered in $g_y=2.2$, is ascribed to a hemoprotein in the low spin state. The principal values, g_x, g_y and g_z, in a g tensor of heme in the low spin state (1.91, 2.25 and 2.41, respectively, in the case of microsomal Fe_x) are close together if $E_\xi - E_\zeta$ and $E_\eta - E_\zeta$ are large, where E_ζ, E_η and E_ξ represent the orbital energies of d_{xy}, d_{yz} and d_{zx}, respectively.[5,6] These matters will be dealt with in more detail in section 3.3.2.

The concentration of heme iron is given by double integration of the differential curve on the recorder of an EPR spectroscope. If the triplet g values are close together, then the signal heights are favorable at the definite concentration of hemoproteins. This is the reason why only the EPR signal of P-450 was observed in the microsomes, although cytochrome b_5 was present at almost the same order of concentration.

From the g values, temperature dependency of the signals, and behavior in oxidation and reduction, it was concluded that the set of triplet g values belonged to a new hemoprotein in the low spin state. Experiments were then carried out to find the natural electron donor to the hemoprotein. Incomplete reduction was observed adding NADH, and reduction to a higher extent was observed with NADPH. In the presence of CO, the hemoprotein was rapidly and completely reduced (Fig. 3.25).[7] From the latter fact, it appeared that the microsomal Fe_x must be related to the hemoprotein, P-450, discovered by Omura and Sato,[1] although doubt was still being cast as to whether P-450 might perhaps be a degradation product of cytochrome b_5. In the case of the EPR observations, however, the non-identity of microsomal Fe_x and cytochrome b_5 was clear, because no EPR signal ascribable to cytochrome b_5 in the microsomes was observable at liquid nitrogen temperature. The latter signal was detectable only when purified cytochrome b_5 was used at a much higher concentration, quite beyond the concentration limit for microsomes condensed as far as possible. Moreover, the g values of cytochrome b_5 are different from those of P-450, being 3.03, 2.23 and 1.43, respectively.[8]

The amount of microsomal Fe_x was expressed as nmole per mg of

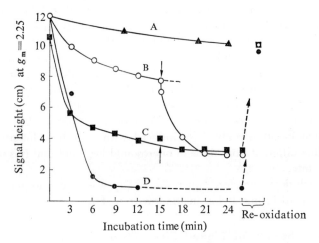

Fig. 3.25. Changes in height of the microsomal Fe_x signal at $g=2.25$, after addition of NADH and NADPH under anaerobic conditions. A, Control; B, NADH added at zero time, then NADPH added at the time of the arrow; C, NADPH added at zero time, then another amount of 0.5 μmole NADPH readded at the time of the arrow; D, NADH added under CO.
(Source: ref. 7 Reproduced by kind permission of the American Society of Biological Chemists, Inc.,)

microsomal protein, when the double integration of the signal was compared under identical conditions with that for a standard solution containing $CuSO_4$ and EDTA. The estimated amount of microsomal Fe_x was about the same as the amount of P-450 determined by spectroscopy, although the accuracy of the EPR determination was inferior to that made by spectroscopy in the case of quantitative estimates. For a rough estimation, the signal heights at $g=2.25$ were used instead of the double integrations. Complete parallelism was found between 450 nm absorbancy and EPR signal height, as far as various rabbit tissues such as liver, adrenal gland, kidney, ovary, etc. were concerned, and the parallelism also existed when the rabbit was replaced by other animals. Moreover, it could even be traced through the course of increasing cytochrome content in liver microsomes during the postnatal growth of rabbits.[9] The parallelism still obtained after the administration of xenobiotics such as phenobarbital[10] or Sudan III.[9] The EPR signal of P-450 which had been purified from *Pseudomonas putida* cultured in the presence of camphor, was found to give similar g values.[11] Consequently, microsomal Fe_x was shown to be identical with cytochrome P-450.

Waterman *et al.* have applied EPR measurements together with spectroscopic observations for determining the oxidation-reduction potential of microsomal Fe_x. The value given was as low as –400 mV at pH 7.0,[12] whereas that of CO-P-450 (Fe^{II})/P-450 (Fe^{III}) has been given as -285 mV at pH 7.0, based on optical measurements[13].

Cytochrome P-450 changes to P-420 when the cytochrome is treated with cholate or other detergents, as observed by spectroscopy of the reduced form under carbon monoxide. Two species of P-420, in different proportions, are produced according to the different treatments, as mentioned below in sect. 3.3.3. One species retains the EPR signal of microsomal Fe_x whereas the other loses it. Moreover, Peisach et al.[14] have reported that P-450 in rabbit liver induced with 3-methylchlolanthrene gives g values different from those of microsomal Fe_x, and that it is in the high spin state.

In conclusion, therefore, microsomal Fe_x was found to represent a species of cytochrome P-450 in the low spin state.

3.3.2 Short Survey of Ligand Field Theory

The spin state, which is discussed generally in this section, is controlled principally by the behavior of electrons in the $3d$ orbitals of heme iron. Although these orbitals do interact with the π and σ orbitals of porphyrin in the heme structure, the coefficients of the $3d$ orbitals are fairly large in comparison with those of the π and σ orbitals. In an approximate evaluation of the magnetic properties of hemoproteins, the delocalization of the $3d$ orbitals to porphyrin may therefore be neglected,[15] and the spin quantum number S is then half-integral and integral in ferric and ferrous hemes, respectively. Depending on the degree of spin pairing of these $3d$ electrons, ferric hemoproteins may assume the electronic structure of iron having five, three, and one unpaired electrons ($S=5/2$, 3/2, and 1/2, respectively), whereas ferrous hemoproteins may form the electronic structure of iron having four, two, and no unpaired electrons ($S=2$, 1, and 0, respectively), as indicated in Fig. 3.26.[16] The first assignment of the spin value S in hemoproteins and their derivatives was made by Pauling and Coryell in 1936,[17] who measured the magnetic susceptibilities of hemoglobin derivatives and demonstrated the dependence of the spin state of hemoglobin on the chemical nature of the 6th ligand. They termed hemoglobin derivatives having $S=5/2$ or $S=2$, *ionic* compounds, and those having $S=1/2$ or $S=0$, *covalent*. According to more recent terminology, these are known as *high spin* and *low spin* compounds, respectively.

Pauling and Coryell[17] found ferric hemoglobin hydroxide to have a magnetic moment of 4.47, which is intermediate between the 5.92 of a purely high spin compound and 1.73 of a purely low spin compound. Since this intermediate value was considered to be characteristic of a

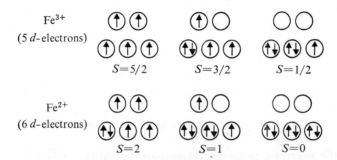

Fig. 3.26. Electronic configurations of ferric and ferrous systems containing 5 and 6 3d-electrons.[16] It is shown by group theory that the d level is split into two sublevels, called $d\gamma$ and $d\varepsilon$, under the influence of a ligand field of cubic symmetry. $d\gamma$ and $d\varepsilon$ comprise 2 and 3 independent orbitals, respectively. The two upper circles in the diagrams indicate the $d\gamma$ orbitals, and the 3 lower circles the $d\varepsilon$ orbitals. These two levels split further if the symmetry of the ligand field becomes lower.

square planar ferric complex having three unpaired electrons, they suggested that, in this compound, four covalent bonds resonate among six coordination positions of iron in an octahedral configuration. Taube[18] suggested on theoretical grounds that this compound may be an equilibrium mixture of high and low spin isomers rather than a compound having three unpaired electrons. In 1956, Griffith[19] made a ligand field calculation for a regular octahedral complex and demonstrated that the electronic state having three unpaired electrons is energetically unstable in such octahedral complexes as ferric hemoproteins. Thus, when the pairing of unpaired electrons takes place in such complexes to decrease the number of unpaired electrons from five to three, further pairing to reduce the number of unpaired electrons from three to one may occur. Griffith[20] further suggested that ferric hemoproteins having magnetic susceptibilities intermediate between high and low spin values may be a thermal mixture of high and low spin isomers based on the Boltzmann distribution. Subsequently, George and his associates[21,22] showed that the magnetic susceptibilities and optical spectra of some hemoproteins are indeed temperature dependent in accordance with Griffith's prediction.

Ingram and his associates[23,24] investigated the electronic structure of the heme iron and the orientation of the heme in ferric hemoglobin, myoglobin, and their derivatives, by EPR spectroscopy. They demonstrated that EPR spectroscopy carried out at cryogenic temperatures is useful for distinguishing high spin hemoproteins from low spin ones. Mössbauer spectroscopy was applied by Lang and Marshall[25] to hemoglobin derivatives in order to investigate the electronic structure of the heme iron in these compounds.

Nevertheless, several investigators[26–28] noted that certain correlations existed between the optical spectra and spin states of hemoproteins, as these parameters were experimentally measured in various hemoproteins and their derivatives. Although empirical, such correlations between the optical spectra and spin states of hemoproteins are very useful for qualitative estimations of the spin state of hemoprotein derivatives by spectrophotometry.

As the magnetic properties of various derivatives of hemoproteins were measured over a wide range of temperatures by these physical techniques, it was recognized that the temperature-dependent changes in the spin state of hemoproteins, which were initially observed at ambient temperatures by George et al.,[21,22] must also occur at cryogenic temperatures. For example, ferric myoglobin azide was reported to have an intermediate magnetic moment at room temperature.[27]

However, the EPR spectrum of this compound indicated that it was purely a low spin compound below $-196°C$.[4,29,30] Another similar example was observed in the case of cytochrome c peroxidase.[31,32] Although cytochrome c peroxidase was found to be essentially in a high spin state at $20°C$,[30] its EPR spectrum showed that the enzyme was a mixture of high and low spin compounds at $-196°C$.[32] It was thus obvious that the spin state of these hemoproteins changed between room temperature and $-196°C$. However, physical techniques suitable for examining such spin transitions over a wide range of ambient and cryogenic temperatures were not well developed, both the EPR and Mössbauer techniques being unsuitable for quantitative studies of spin transition at or near ambient temperatures, since the relaxation characteristics of the heme iron become unfavorable for these techniques at higher temperatures.

Yonetani et al.[33] demonstrated in 1966 that the optical spectra of cytochrome c peroxidase changed reversibly with temperature below $0°C$. It was qualitatively determined by applying the above-mentioned empirical correlation between the optical and magnetic properties of heme compounds,[21,22,27,28] that the transition of the spin state of cytochrome c peroxidase occurred in a narrow range of cryogenic temperatures between $-100°C$ and $-10°C$.

More recently, Iizuka and Kotani[34,35] have constructed a highly stable magnetic torsion balance which is operational at both ambient and cryogenic temperatures. With this instrument, they have extensively investigated the temperature-dependent transition of the spin state of several hemoproteins and their derivatives over a wide range of temperatures.[34-37] The precise determinations of the magnetic susceptibilities of hemoprotein derivatives carried out by these investigators have made it possible to estimate in detail the thermodynamic parameters involved in such spin transitions, and to speculate on the probable mechanisms of such spin transitions in hemoproteins.

Concerning the temperature dependency of the equilibrium between high and low spin state P-450, no systematic investigations have yet been completed. However, from the optical and EPR data so far known, it appears that the equilibrium may be on the low spin state side in untreated and phenobarbital-treated microsomes from rabbit liver, when substrate is absent.

Apart from the above methods, ligand field theory will undoubtedly prove a very powerful tool in the NMR, magnetic circular dichroism, and resonance Raman spectroscopy of hemoproteins.

We will next deal with the relationship between the g values and orbital energy differences in the d orbitals of heme iron, to estimate the rhombicity of P-450 heme ligands. If five electrons of a heme iron are accommodated in three $d\varepsilon$ orbitals of the low spin form, there are three ways of distributing the electrons: $\xi\eta^2\zeta^2$, $\xi^2\eta\zeta^2$, and $\xi^2\eta^2\zeta$, leaving an electron hole in $\xi(d_{xy})$, $\eta(d_{yz})$, and $\zeta(d_{zx})$, respectively. In a ligand field of cubic symmetry, these three electronic configurations degenerate when making a spin doublet, $^2T_2$. If the symmetry of the field is lower than cubic, then considering the total spin of the state to be half-integer (1/2), we expect the six-fold degenerated spin-orbitals to split into three Kramers' doublets, of which the energies are depicted as E_ξ, E_η, and E_ζ. However, in this time the wave function of the lowest Kramers' doublets will not be represented by a pure hole configuration, ξ, η, or ζ. The spin-orbit interaction tends to mix these three "subconfigurations"; that is,

$$\left.\begin{aligned}\psi^+ &= A\xi\alpha + iB\eta\alpha + C\zeta\beta \\ \psi^- &= -A\xi\beta + iB\eta\beta + C\zeta\alpha\end{aligned}\right\} \quad (3.1)$$

The theoretical g values are obtained from the Zeeman splitting of the ground state $\psi^\pm$ when a static external magnetic field $\boldsymbol{H}$ is applied:[5,6]

$$\left.\begin{aligned}g_x &= 2|A^2 - (B+C)^2| \\ g_y &= 2|-(A-C)^2 + B^2| \\ g_z &= 2|(A-B)^2 + C^2|\end{aligned}\right\}, \quad (3.2)$$

and

$$A^2 + B^2 + C^2 = 1. \quad (3.3)$$

These equations form an over-determined system for the three unknowns A, B, and C. Using the ratio of the experimental g values to calculate a set of A, B and C values from Eq. (3.2) and (3.3), we obtain the following equations

$$A = \frac{-(g'_z + g'_y)}{\sqrt{(g'_z + g'_y)^2 + (g'_z - g'_x)^2 + (g'_y - g'_x)^2}}$$

$$B = \frac{(g'_z - g'_x)}{\sqrt{(g'_z + g'_y)^2 + (g'_z - g'_x)^2 + (g'_y - g'_x)^2}}$$

$$C = \frac{(g'_y - g'_x)}{\sqrt{(g'_z + g'_y)^2 + (g'_z - g'_x)^2 + (g'_y - g'_x)^2}}$$

(3.4)

where g'_x, g'_y and g'_z represent the observed g values.

The values of g_x, g_y and g_z are calculated using Eq. (3.2) for checking the consistency from the A, B and C obtained by Eq. (3.4). Usually, they take almost the same values as g'_x, g'_y and g'_z. If the $A\xi\alpha$ and $-A\xi\beta$ in Eq. (3.1) are unperturbed terms and the others represent the mixing of the excited doublets by perturbation, B/A and C/A can be related to the orbital energy differences between E_ξ, E_η and E_ζ through the well-known formulas of quantum mechanical perturbation theory: the larger the differences, the smaller the value of B/A or C/A. Then the energy differences between the orbitals are given in terms of g's in rather simple form as follows.

$$E_\xi - E_\eta = \frac{g_z + g_y}{g_z - g_x} \cdot \frac{a}{2} \quad \text{and} \quad E_\xi - E_\zeta = \frac{g_z + g_y}{g_y - g_x} \cdot \frac{a}{2},$$

(3.5)

where a is a spin-orbit coupling constant. As the energy differences concern the positive hole, the actual energy differences of the orbitals, which accommodate the electrons, are of reverse sign. In the case of P-450 in the low spin state, taking the sign into consideration,

$$E_\eta - E_\xi = \frac{(2.41 + 2.25)}{(2.41 - 1.91)} \cdot \frac{a}{2} = 4.7a$$

and

$$E_\zeta - E_\xi = \frac{(2.41 + 2.25)}{(2.25 - 1.91)} \cdot \frac{a}{2} = 6.9a.$$

The spin-orbit coupling constant in a complex is usually smaller than a free-ion value owing either to the contamination of higher excited states to the lower, or to reduction of electron density on the central ion due to the formation of molecular orbitals with the ligand. However, it is very difficult to estimate the reducing factor. If a is assumed to be 400 cm^{-1}, which is the free-ion value in Fe^{3+}, the d_{zx} and d_{yz} orbital energies are about 2800 cm^{-1} and 1900 cm^{-1}, respectively, above that of the lowest d_{xy} orbit.

Next, we will discuss the high spin case, $S = 5/2$. The calculations are completely parallel to the low spin case, the ground state being $^6A_1$ $(d\varepsilon^3(d^1_{xy}, d^1_{yz}, d^1_{zx})d\gamma^2)$.

In an orthorhombic field, the lowest excited state $^4T_1$ splits into the doubly degenerated quartet $^4E'(d\varepsilon^4(d^1_{xy}, d^2_{yz}, d^1_{zx})d\gamma^1$ and $d\varepsilon^4(d^1_{xy}, d^1_{yz}, d^2_{zx})d\gamma^1)$ and the orbitally non-degenerated quartet $^4A'_2$, where the z axis is taken to coincide with the direction of orthorhombic elongation or compression. Since the two states belonging to $^4E'$ transform similarly to the pair (x,y) under rotation of the coordinates, we depict their energies as E_x and E_y. Thus, the perturbation by spin-orbit interaction mixes the lower excited states of energies E_x, E_y and E_z to the ground state $^6A_1$, which splits into three Kramers' doublets corresponding to the values $\pm 1/2$, $\pm 3/2$ and $\pm 5/2$ for the spin components along the z axis.

This splitting scheme can be summarized as a spin-Hamiltonian, in which the effects of spin-orbit perturbation are represented by the two parameters, D and E.[38]

$$\mathscr{H} = 2\beta \boldsymbol{H}\cdot\boldsymbol{S} + D\left(S_z^2 - \frac{1}{3}S(S+1)\right) + E(S_x^2 - S_y^2) \quad (3.6)$$

$$\left.\begin{array}{l} D = \dfrac{a^2}{10}\left(\dfrac{2}{E_z - E(^6A_1)} - \dfrac{1}{E_x - E(^6A_1)} - \dfrac{1}{E_y - E(^6A_1)}\right) \\[2mm] E = \dfrac{a^2}{10}\left(\dfrac{1}{E_x - E(^6A_1)} - \dfrac{1}{E_y - E(^6A_1)}\right) \end{array}\right\} \quad (3.7)$$

The relationship between the g values and orbital energy differences in the high spin state is deduced by diagonalization of the spin-Hamiltonian with $S=5/2$. Assuming that $E \ll D$ and treating the E term as a small perturbation, the second-order calculation of the splitting of g_x and g_y in a rhombic ligand field gives

$$g_x = 6 + 24 E/D, \quad g_y = 6 - 24 E/D. \quad (3.8)$$

If the E/D value becomes larger, that is, the symmetricity is reduced, then the deviation of the g_x and g_y values from 6 becomes large.

Using experimental values of g_x and g_y, it is possible to estimate the E value, if D is assumed to be 10 cm^{-1}.[39] Then, if the energy of $^4E'$ before separation is assumed to be 6000 cm^{-1} above that of the $^6A_1$ state, the energy difference between E_x and E_y can be evaluated. In the case of P-450, this value is estimated to be about 1800 cm^{-1}.

3.3.3 High Spin P-450 Induced by 3-Methylcholanthrene

In the preceding section, the high spin form of P-450 was introduced without showing the shape or intensity of the EPR signal. Peisach and Blumberg[14] have demonstrated a high spin EPR spectrum of P-450 from 3-methylcholanthrene-induced rat liver microsomes. The g values are quite different from the usual high spin g values of several hemo-

116 MOLECULAR PROPERTIES

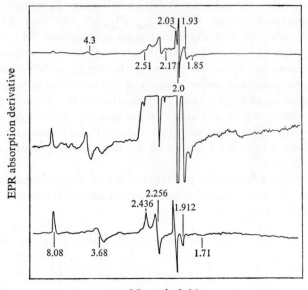

Fig. 3.27. EPR absorption derivatives spectra of a thick slice of rabbit liver (upper and middle curves) and rabbit liver microsomes (low curve) from a methylcholanthrene-treated animal. The spectra were all taken at 1.60 K. The gains were increased by a factor of 10 for the middle trace and by a factor of 5 for the lower trace. The g values of various features of the spectra are indicated on the figure.
(Source: ref. 14. Reproduced by kind permission of the National Academy of Sciences, U.S.A.)

proteins: that is to say, they are near 8, 3.7 and 1.7 (Fig. 3.27). They are detected only at liquid helium or liquid hydrogen temperature. From ligand field theory, these values reveal that the axial symmetry is also lost in this high spin form of P-450 and a high rhombicity is present as stated in section 3.3.2.

The high spin signal was not due to complex formation between low spin P-450 and 3-methylcholanthrene (3-MC), since neither an increase in the high spin signal was observed by adding 3-MC to untreated microsomes nor was the inducer 3-MC, when ^{14}C-labeled, found in the preparation for EPR experiments.[40] When actinomycin D was administered to rabbits together with 3-MC, no increase in high spin P-450 was seen and the content of P-450 in the low spin state remained at normal or just below normal levels.[41] The above-mentioned high spin P-450 is induced by carcinogens such as 3-MC and benzo(a)pyrene or 2,3,7,8-tetrachlorodibenzo-p-dioxin (TCDD), a contaminant from herbicide production and a potent aryl hydrocarbon hydroxylase inducer.[42] The 3-MC induction of aryl hydrocarbon hydroxylase, 7-ethoxycoumarin O-deethylase, p-nitroanisole O-demethylase, and 3-methyl-4-methylaminoazobenzene N-demethylase activity is associated with the same genetic region. Genetically "responsive" strains of mice show induction of the monooxygenases and concomitant new formation of cytochrome P_1-450, which possesses a blue-shifted peak at 448 nm in the difference spectrum of the CO-reduced form. "Non-responsive" strains for 3-MC or benzo(a)pyrene appear to require more TCDD to induce these characteristics. The cytochrome P_1-450 shows the high spin EPR signal as reported by Nebert et al.[42,43]

Cytochrome P-450 from *Rhizobium japonicum* has been fractionated into components which were almost completely high spin ($g=7.9$, 3.8 and 1.75, P-450b_1) and almost completely low spin ($g=2.42$, 2.25 and 1.92, P-450a; $g=2.40$, 2.24 and 1.92, P-450b; $g=2.43$, 2.26 and 1.92, P-450c)[45]

Denaturation of P-450 does not bring about the rhombic EPR high spin signal. As Mason et al.[46] reported, PCMB-degraded P-450 was found to show g_x, $g_y=6$, which is the usual g value for high spin heme. The deviation from $g=6$ is less than 10% in abnormal methemoglobin[47].

Although the aryl hydrocarbon hydroxylase system, which seems to involve rhombic high spin P-450, can utilize NADH as the electron donor for 3-MC hydroxylation, it remains to be determined whether the NADH-dependent P-450 reported by Ichikawa and Loehr[48] is associated with the former P-450.

According to the estimation of Nebert et al.,[42] the quantity of rhombic high spin P-450 in 3-MC-treated C57BL/6N mice was much larger than that of low spin P-450, about 4/5 being in the high spin form. The determination was based on double integration of the recorded EPR signals. The content of high spin P-450 is very large, when the signal heights are comparable between the high and low spin states, since the integration range is wide for the high spin form from $g=8.01$ to $g=1.74$. They also stated that about 1/2 of the P-450 was in the high spin form in control and phenobarbital-treated mice at the same temperature (about 10 K). These results are inconsistent with those obtained from the EPR and absorption of rabbit liver microsomes at liquid nitrogen temperature. The following points are important for resolving this matter: (1) measurements should be carried out with microsomes from different animal sources, (2) the equilibrium might be shifted at liquid helium temperature from low spin to high spin, although this is not usually the case, (3) as for the EPR power saturation, the signal of low spin species seems to be easily saturated at liquid helium temperature, when compared to the signal of high spin species, so that a saturation effect might lead to large errors in the ratio of the contents of high spin to low spin species, and (4) the content determination by double integration on the recorded EPR curve is not easy when the EPR is observed in a wide range of magnetic field, since the transition probability between the Kramers' doublet which depends on the g values must be taken into consideration.[49]

3.3.4 EPR Signal Changes of P-450 with Various Ligands or Substrates

Before discussing the interaction of P-450 with small molecules or substrates, we will deal briefly with the coordination of the 5th or 6th position of the heme.

The suggestion was first put forward by Murakami and Mason[46] that the sulfhydryl group of P-450 could be associated with the EPR signal of microsomal Fe_x because the addition of p-chloromercuribenzoate (PCMB) diminished the EPR signal. In 1966, Hollocher[50] described the EPR signal changes occurring in the course of alkali titration of methemoglobin. At the alkaline side, over pH 11, a triplet set with g values of 2.41, 2.25, and 1.91 was observed. He did not, however, describe any assignment for the EPR signal. We have also noticed that these signals disappear on addition of PCMB, but reappear on introduction of glutathione, as in the case of the reversible action of PCMB on P-450.[51] The g values of the EPR for the interaction of low molecular heme (as well as hemoproteins) with mercaptans were reported by Bayer et al. in 1969.[52] Model experiments were also carried out independently by Jefcoate and Gaylor[53] with metmyoglobin and n-propylmercaptan. A series of hydrophobic sulfhydryl compounds were prepared by us by cleavage of the -S-S- bond of the compounds listed in Table 3.10.[54] When these compounds coexisted with metmyoglobin, EPR signals within a limited range (as listed in the table) appeared instantaneously, whereas metmyoglobin itself did not interact with glutathione for several hours. When these hydrophobic sulfhydryl compounds were reacted with P-450, the microsomal Fe_x signal was replaced by signals similar to those of metmyoglobin-SH compound complexes.[54] Concerning ferric cysteine and heme cysteine interaction, Bayer et al.[55] have given a more precise description including their experimental results and certain theoretical aspects. Although there is a good similarity in EPR signals between P-450 and the metmyo(hemo)globin-heme-S⁻ complex, the absorption of the CO-complex of the reduced form is distinctly different: the model system shows a usual absorption of the low spin state, viz. a peak at around 420 nm. Recently, cytochrome P-450-like absorp-

TABLE 3.10. g Values of methemoglobin-thioalcohol complexes

	g_x	g_y	g_z
TED†	1.91	2.23	2.37
TOED	1.92	2.24	2.37
TAD	1.91	2.24	2.37
TBD	1.91	2.25	2.24
Microsomal Fe_x	1.92	2.25	2.42
TAD-treated microsomes	1.93	2.25	2.41
	1.95	2.23	2.37

† Thiamine disulfide derivatives:

R	
$-CH_2CH_3$	TED
$-CH_2CH_2OH$	TOED
$-CH_2CH=CH_2$	TAD
$-CH_2CH_2CH_2CH_3$	TBD

(Source: ref. 54. Reproduced by kind permission of University Park Press, U. S. A.)

tion has been obtained by Stern and Peisach[56] in a model system composed of ferrous heme, CO, thiol and a strong base in a dimethylsulfoxide-ethanol solvent. Another model experiment for the high rhombic EPR signal of high and low spin ferric heme has been presented by Collman et al.,[57,58] using meso-tetraphenylporphyrin and benzenethiol. Consequently, one of the axial ligands may be a $-S^-$ anion of a cysteinyl residue.

The candidate for the other axial ligand is the histidyl residue of P-450 apo-protein. This is suggested by the dependency of the CO-binding strength of reduced P-450, the inflexion point at around pH 7.5[59] Another suggestion is given by the EPR signal of the complex of metmyoglobin with a sulfhydryl compound, in which a histidyl residue may be one of the axial ligands. Dus et al.[60] have also proposed the same idea based on the fact that the α- and β-bands of oxidized P-450$_{CAM}$ were found to be shifted to shorter wavelengths by the substrate and that N-phenylimidazole reversed this effect.

As described elsewhere, the EPR signals of hemoproteins are dependent mainly on the ligand states of heme iron. The EPR signal of normal P-450 in the low spin state can be modified by binding to added compounds which are low spin ligands or high spin ligands. Observations on the EPR signal changes of P-450 may thus give direct evidence on the changes in heme ligands, and we can discuss the nature of the binding of a ligand, binding strength or interaction of a ligand with the neighboring residues, from EPR experiments.

As shown in Table 3.11,[59] cyanide, hydroxide and isocyanides

TABLE 3.11. g Values and orbital energies of P-450[59]

	Fe_x[†1]	Fe_x-CN[†2]	Fe_x-OH	Fe_x-EtNC[†1]
g_x	1.91 (1.90)[†3]	1.85 (1.83)	1.86 (1.86)	1.87 (1.87)
g_y	2.25 (2.24)	2.32 (2.30)	2.25 (2.26)	2.29 (2.29)
g_z	2.41 (2.40)	2.53 (2.51)	2.48 (2.48)	2.43 (2.43)
$E_\xi - E_\eta$	4.66 a [†4]	3.57 a	3.82 a	4.21 a
$E_\xi - E_\zeta$	6.85 a	5.16 a	6.06 a	5.62 a
$E_\xi - E_\eta$[†5]	4.06 a	3.43 a	5.90 a	6.56 a
$E_\xi - E_\zeta$	8.03 a	5.03 a	8.11 a	7.57 a

[†1] Fe_x, microsomal Fe_x; EtNC, ethylisocyanide.
[†2] Similar g values have also been given by Schleyer et al.[61] and Gunsalus et al.[62]
[†3] Theoretical values, calculated by Kotani's method.
[†4] Spin-orbit coupling constant.
[†5] According to Griffith's method.

bind to P-450, changing the g values from the native ones. In Table 3.11, the orbital energy differences are also listed, as calculated by the procedure mentioned above. The rhombicity is decreased by the binding of P-450 to these small molecules. The binding of isocyanides to P-450 has been studied in detail by us. The EPR signals of methyl-, ethyl- and *tert*-butylisocyanide-P-450 complexes were almost identical, showing g values of g_x=1.88–1.89, g_y=2.28–2.29 and g_z=2.43, while the EPR signal of phenylisocyanide-P-450 complex formed a set of perturbed values, g_x=1.89, g_y=2.28 and g_z=2.40, representing the interaction of phenyl residue of the isocyanide with the hydrophobic portion of P-450 in the neighborhood of the heme. The binding strength increased in the above-mentioned sequence, showing that hydrophobic interaction plays a role in complex formation.[63]

In addition to these EPR studies, the same sequence was obtained by spectrophotometric measurements, in which the peak at 434 nm and the trough at 400 nm were recorded. Theoretically, EPR signals of reduced P-450 are undetectable in either the high spin state or the low spin state. Observations on reduced isocyanide-P-450 complexes revealed the same sequence in binding strength as in the oxidized P-450 complexes, but with the affinity two orders of magnitude higher in the former than in the latter. As a consequence of the increased affinity of P-450 in the reduced form for isocyanides, oxygen could hardly replace the ligand and auto-oxidizability disappeared in the isocyanide-P-450 complex.[63] The binding of spin-labeled isocyanide to P-450 has been reported by Reichman et al.[64] By addition of cyanide at a high concentration such as 20 mM, the EPR signal of P-450 decreases to less than half with the appearance of the signal of the P-450-cyanide complex.

The affinity of P-450 in rabbit liver microsomes is likely to be weak for cyanide. However, the occurrence of P-450 species in rat liver microsomes showing very strong affinity has been reported by Gaylor et al.[65] The EPR signal changes of P-450 in adrenocortex and in Pseudomonos have also been reported by Schleyer et al.[61] and Gunsalus et al.,[62] respectively.

Nitric oxide is frequently used to search for the character of heme as well as mobilizability of bound NO in the crevices of hemo protcins, in order to obtain information on the mechanism of oxygenation of oxygen activation of the hemoproteins. Nitric oxide could be a spin-labeled model compound of an oxygen molecule.

Miyake et al.[66] first reported the EPR signal of a well-separated triplet due to the nitrogen nuclear magnetic moment with both oxidized and reduced P-450 from microsomes and submicrosomal particles. With purified $P-450_{CAM}$, microsomal P-450 and microsomal P-420, EPR measurements of P-450 (or P-420)-NO complexes were carried out by Ebel et al.[67] According to their experiments, both $P-450_{CAM}$ and microsomal P-450-NO complexes in the ferric state lacked EPR signals and the ferrous microsomal P-450-NO complex was unstable. The latter complex was converted to a species which had an EPR spectrum similar to that of P-420-NO. Hyperfine splitting for the g_z signal of 20 G was observed in the unstable microsomal P-450-NO complex. The EPR signal of the P-420-NO complex with narrower hyperfine splitting than that of the P-450-NO complex resembled that of sodium dodecyl sulfate-modified hemoglobin.[68,69]

Peisach et al.[70] have presented information on the relationship between different ligands and the g values of the lowest field (Table 3.12).

TABLE 3.12. Typical low-field g values of hemoprotein compounds where one of the z ligands is a histidine imidazole

Ligand	Amine	Met	Imid	Imid⁻	OH⁻	RS⁻
g_x	3.3	3.15	3.0	2.8	2.5	2.4

(Source: ref. 70. Reproduced by kind permission of Williams & Wilkins Co., U.S.A.)

Also, information obtained concerning the axial ligands of P-450 is helpful for analyzing the results of substrate binding, i.e. whether the substrate binds directly as ligand or perturbs the g values by binding in the neighborhood of the ligand.

There are numerous reports on the spectral changes ascribed to the binding of substrates to P-450. If a substrate occupies the 5th or 6th coordinate of P-450 or dislocates the ligand from the native position,

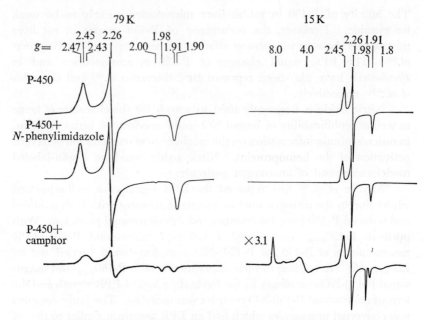

Fig. 3.28. P-450$_{CAM}$ EPR spectra at 15°K and 79°K. Top curve, P-450$_{CAM}$ 1.1 mM, in 50 mM phosphate buffer, pH 7.4; middle curve, plus 1.5 mM N-phenylimidazole; bottom curve, plus 1.5 mM (D+)-camphor.
(Source: ref. 71. Reproduced by kind permission of the National Academy of Sciences, U.S.A.)

then the EPR signal can be expected to change. The latter case is seen with P-450$_{CAM}$ from *Pseudomonas putida* cultured in the presence of camphor, which shows an EPR signal change from the low spin form to the high spin form on camphor addition. As shown in Fig. 3.28,[71] the EPR signal at g=2.45, 2.26 and 1.91 decreased upon addition of camphor (the bottom curves) and a new high spin signal appeared at 15 K. The g values were 8.0, 4.0 and 1.8 (diffuse). In the middle curves, the EPR signal of P-450 complexed with phenylimidazole (which is a specific nitrogenous effector, reacts at 10^{-3}M and displaces substrate) is shown. The g values of the complex are slightly shifted from the native ones, at g=2.47, 2.26 and 1.90.

In contrast, the EPR signal changes upon addition of substrate from low spin form to high spin form or vice versa have remained ambiguous with liver microsomal P-450. Although the binding of substrate to liver P-450 has been established on the basis of difference absorption spectra, the spectral changes are very small compared to those of bacterial P-450. That is to say, the difference millimolar extinction coefficient of the so-called Type I or Type II spectrum is less than 20 mM^{-1}

·cm^{-1}, as described in section 3.2. Such absorption spectral behavior is reflected in the EPR signal changes. An EPR signal change in P-450 in the presence of substrate is not always apparent, even when a distinct difference spectrum is observed.[72] The complexity of the relationship between the EPR and absorption spectral changes may be attributable to the multiplicity of liver microsomal P-450 and to the fact that one of the components converts its spin form from the low spin to the high spin form or vice versa while the other P-450's remain in the same spin state.

Such complexity was also seen in experiments concerning the EPR signals and the absorption spectra of liver microsomes (Fig. 3.29).[73] The low spin region EPR spectrum before the addition of ethanol and

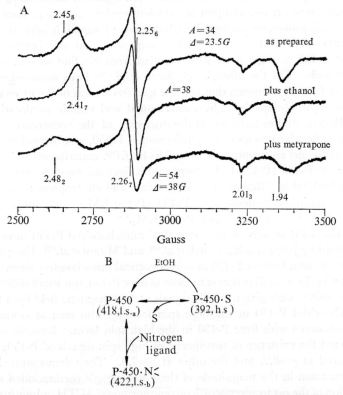

Fig. 3.29. A, EPR spectrum of low spin cytochrome P-450 from liver microsomes in the presence of ethanol and metyrapone. B, A generalized scheme, based on the absorption spectra, showing the possible interconversions of ferric P-450 from liver microsomes. The wavelengths indicated refer to the absorbance band maxima observed in the absolute absorption spectrum. Two low spin (l.s.) forms of P-450 and a high spin (h.s.) form are indicated. (Source: ref. 73. Reproduced by kind permission of Williams and Wilkins Co., U.S.A.)

metyrapone (Fig. 3.29A) does not seem reconcile with data obtained from the changes in the absorption spectrum (Fig. 3.29B), where high spin form of P-450 is shown before the addition of ethanol. This complication can be attributed to the existence of multiple forms of P-450.

EPR spectra of P-450 complexed with aniline have also been given by Estabrook et al.[74] They reported that the triplet set of g values, 2.42, 2.26 and 1.92, was converted to 2.45, 2.26 and 1.90 upon addition of 5 mM aniline, a sufficiently high concentration to be coordinated to the heme iron. Waterman et al.[75] have reported a heterogeneity in the EPR signal of microsomal Fe_x, which was modified to give a decreased low spin signal with an increased high spin signal in about 30% of the original heme on the addition of sufficient cyclohexane, and then to an apparently homogeneous signal with recovered signal height after the addition of 30 mM hexanol. They suggested that the heterogeneity of the P-450 as regards its EPR resulted from binding to endogenous compounds. On the other hand, Stern et al.[76] have reported on the basis of EPR experiments that the low spin ferric form of P-450 consisted of more than one species in liver microsomal and soluble preparations.

Recent studies have led to the discovery of the occurrence of different molecular species of liver microsomal P-450, as observed by polyacrylamide gel electrophoresis as well as by EPR experiments.[77-79] Cytochrome P-450 (P-450 LM_2) from rabbit liver microsomes induced by phenobarbital is different in its EPR signal and electrophoretic mobility from β-naphthoflavone-induced P-450 (P-450 LM_4), and antibodies to P-450 LM_2 do not react with P-450 LM_4.[80]

The EPR signals of adrenocortical mitochondrial P-450 have been described by Jefcoate et al.,[81] Bell et al.[82] and Mitani et al.[83] The g values of the low spin form of P-450 in adrenocortical mitochondria were observed to be in a similar region to those of liver P-450, but small differences in the values were given. EPR signals at low magnetic field for adrenal mitochondrial P-450 in the high spin form are also seen at around $g=$ 8, as observed with liver P-450 in the high spin form. Brownie et al.[84] proposed the existence of two species of high spin signals of P-450: one appeared at $g=8.2$, and the other at $g=7.9$. They demonstrated that the variation in the magnitude of these two signals corresponded to the reaction of the rat to adrenocorticotropic hormone (ACTH) administration or to the stress applied to the animal. They suggested on the basis of results showing a good correlation between the Type II spectral changes and the magnitude of the $g=8.2$ signal, that the $g=8.2$ signal is produced by high spin P-450$_{scc}$ associated with cholesterol side-chain cleavage and the $g=7.9$ signal is produced by high spin 11 β-hydroxylase cytochrome

P-450. This conclusion was confirmed in more detail later by Jefcoate et al.[85] and Brownie's group,[86] although the former gave g values of 8.2 and 8.1 for high spin signals of P-450$_{scc}$ and 11β-hydroxylase P-450, respectively.

The binding of steroids to P-450 changes the EPR spectral pattern; that is, it changes the spin state of the heme of the P-450. In the case of P-450$_{scc}$ preparations,[81,83] which represent high activity for cholesterol side-chain cleavage, relatively weak low spin signals ($g=2.43$, 2.25 and 1.91) were observed. This can be explained on the basis of the fact that endogenous cholesterol may cause a shift of spin state from low to high, as suggested by various reports.[87,88] For example, the action of cholesterol oxidase on the P-450 preparation changed its Soret absorption from the high spin type to the low spin type.[88]

In contrast to cholesterol, 20α-hydroxycholesterol has been found to intensify low spin signals at $g=2.4$, 2.2 and 1.9.[81,83] It remains to be determined whether the binding of 20α-hydroxycholesterol influences the P-450 as a substrate, or an alcohol or detergent. The addition of alcohols or detergents caused changes in the EPR signals of P-450's from both adrenal mitochondria and liver microsomes.[88-90]

3.3.5 EPR Study of P-450 Conversion to P-420

Cytochrome P-450 is easily converted to P-420 under various conditions. The latter species is designated by spectroscopic observations, and shows a peak at 420 nm on reduction with dithionite under CO. The conversion is brought about by many reagents such as detergents and mercurials.

Parallelism between the optical absorbancy and EPR signal height does not always occur when P-450 is converted to P-420. If liver microsomes are treated with 4% deoxycholate for 10 min at 15°C, the peak at 450 nm of reduced P-450 under CO is replaced by the 420 nm peak, while the EPR signal remains unchanged, still showing the low spin signal of microsomal Fe$_x$ (Fig. 3.30).[91] After the addition of 20% glycerol to the reaction mixture, the optical absorbancy at 450 nm is recovered and parallelism reappears. After treatment of microsomes with urea, amides, dioxane, sodium dodecyl sulfate, guanidine salts or phospholipase, the P-420 obtained loses the EPR signal of microsomal Fe$_x$ and cannot be reconverted to P-450 by polyols or glutathione. The presence of glycerol prevents P-450 from conversion to P-420 and the P-450 remains intact at pH 7.0 for more than 8 days at 16°C even under aerobic conditions. The change from P-450 to P-420 occurs with mercurials but glutathione reverses the reaction. The EPR signal with PCMB shows

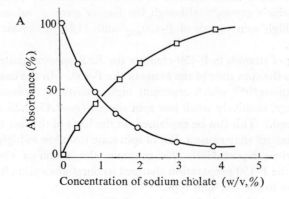

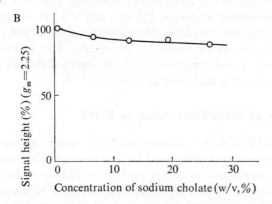

Fig. 3.30. A, Relationship between microsomal CO-binding P-450 and P-420 at various concentrations of sodium cholate. The test samples were incubated at 15°C for 10 min with various sodium chlolate concentrations. The ordinate shows the percentage of the absorbance between 450 and 490 nm for P-450, or the absorbance between 420 and 490 nm for P-420. ---, P-450; ---, P-420. B, Effect of sodium cholate on microsomal Fe_x. The test samples were incubated at 15°C for 10 min with various sodium cholate concentrations. The height of the EPR signals of microsomal Fe_x was measured at $g=2.25$. The points show the percentage of the height of the initial EPR signal.

a high spin signal of $g=6.0$, as mentioned above.

Subparticles having the EPR signal of microsomal Fe_x but containing P-420 rather than P-450 based on their absorbancy, have been prepared by treatment of microsomes with trypsin.[92] Thus, another factor other than the axial ligands of the heme may contribute to the appearance of the 450 nm peak of CO-reduced P-450.

3.3.6 NMR Study of P-450

High resolution nuclear magnetic resonance (NMR) spectroscopy of paramagnetic macromolecules has recently been developed as a powerful tool for investigating structure and structure-function relationships in metalloproteins such as hemoproteins and iron-sulfur proteins. It has been demonstrated that hyperfine shifted resonances represent sensitive probes for elucidating the environment of the metal ion, e.g. in detecting cooperativity in hemoglobins, elucidating the tertiary structure of myoglobin, and characterizing the ligand binding and configuration around the heme iron. This is considerably advantageous when studying metal-containing proteins.[92-102]

In the case of ferric hemoproteins, most investigators have studied purely low spin compounds such as cyanide complexes of myoglobin and hemoglobin, cytochromes c, b_5 and b_2, etc. In these cases, it is difficult to extract clear information from the NMR spectra of the protein itself (in the aromatic and aliphatic regions), but the proton signals of the heme side chain and ligands are readily observed due to their large and appropriate hyperfine shifts arising from ferric low spin iron ($S=1/2$).[92-102]

Next, NMR results for various low spin compounds of ferric cytochrome P-450$_{CAM}$ will be introduced and compared with data for other ferric hemoproteins.[102] Fig. 3.31A illustrates the downfield side of 220 MHz proton NMR spectra of cyanide complex of ferric cytochrome P-450$_{CAM}$ as well as of myoglobin cyanide and horse radish peroxidase cyanide complexes. As can be seen, the signals derived from the cyanide form of cytochrome P-450 were detectable from a single scan at 21°C. After time average accumulation, the general features of the spectra including the position, intensity and linewidth of the signals, became more prominent. When the temperature of the sample was varied between 21 and 45°C the signals at both -23.0 and -14.7 ppm were found to shift in accordance with Curie's law, while another signal at -12.0 ppm did not. Thus, the former two peaks were assignable to protons near the paramagnetic center. This fact, together with relative intensities of the peaks, indicated that the signals at -23.0 and -14.7 ppm were due to methyl and presumably vinyl protons at the side chain of protoheme, respectively. The origin of the third peak at -12.0 ppm, which did not obey Curie's law with varying temperature, is unknown at present. Likewise, the reason for the relative broadness of the former two peaks is not yet clear; it could be due either to restricted rotation of these groups by strong van der Waals contacts with apo-protein or to the high viscosity

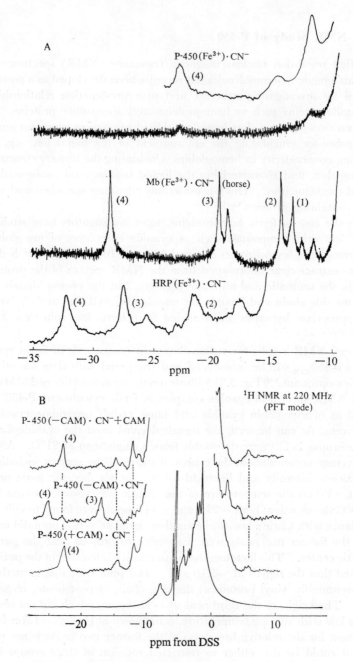

Fig. 3.31 (Legends see p. 129)

of the sample solution.

Another feature of Fig. 3.31A is that the methyl peaks of cytochrome P-450$_{CAM}$ are located at higher fields than those of the corresponding derivatives of other hemoproteins such as myoglobin and horseradish peroxidase. No signal was detectable at fields lower than -23.0 ppm with the cyanide complex of the cytochrome, despite extensive efforts using time-average accumulation over fields down to -50 ppm. Thus, the peak at -23.0 ppm represents the signal lying at the lowest field among the four protoheme methyl signals. The other three must be obscured under the bulky nonspecific signals at higher fields, indicating that the hyperfine shifts of methyl protons of the heme side chain are smaller in the case of cytochrome P-450$_{CAM}$ than with other hemoproteins. The contact shift, which is related to the electron spin density distributed on the proton, has been shown to dominate the hyperfine shift of the methyl protons in the heme side chain of myoglobin complexes.[103] These results, therefore, can be taken as proof that the odd electron density of the heme in cytochrome P-450$_{CAM}$ is low at the periphery of the porphyrin. This, in turn, implies a localized spin density at the center.

Fig. 3.31B illustrates the effect of substrate (camphor) on the hyperfine shifted signals of a ferric cyanide complex of cytochrome P-450$_{CAM}$. In the absence of substrate, two methyl signals (4) and (3) appear in the lower field, whereas in the presence of substrate only one methyl signal (4) is observed in the higher field and other methyl signals are hidden in the protein region. This leads to two conclusions as follows.

(1) NMR can detect the difference between binary (enzyme-ligand) and ternary (enzyme-ligand-substrate) complexes.

(2) The substrate (camphor) induces a change in the electron dis-

Fig. 3.31. A, Down field side of proton NMR spectra of cyanide complex of ferric cytochrome P-450$_{CAM}$, compared with those of horse myoglobin and horse radish peroxidase.[16] Measurements were carried out with a Varian Associates HR-220 high-resolution spectrometer (Dept. of Hydrocarbon Chemistry, Faculty of Engineering, Kyoto Univ.) operating at 220 MHz with a standard variable temperature unit. The temperature was kept at 21°C. The resonance positions were determined with respect to the external reference of 2,2,3,3-tetradeutero-3-(trimethylsilyl)propionic acid sodium salt. In some experiments delicate signals were improved by the use of a Varian Associates c-1024 computer of average transients. (1)–(4) indicate signals from the methyl group of the heme ring.

B, Effect of substrate (camphor) on the NMR spectra of the cyanide complex of ferric cytochrome P-450$_{CAM}$. The lower and middle spectra represented by P-450(+CAM)·CN$^-$ and P-450(−CAM)·CN$^-$, were observed in the presence and absence of substrate, respectively. The upper spectrum (P-450(−CAM·CN$^-$ +CAM) gives the NMR signal measured just after the addition of substrate to the sample in the middle spectrum, which reproduces the bottom spectrum. The experimental conditions were the same as in A, except that a Nicolet Pulsed Fourier Transform unit TT-100 was utilized.

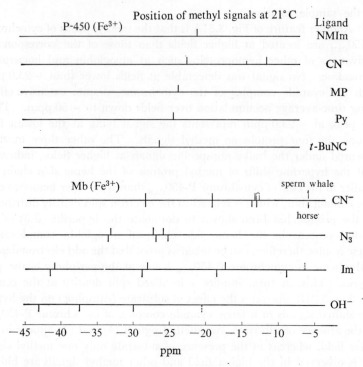

Fig. 3.32. Stick diagram indicating the position of each methyl signal for N-methylimidazole (NMIm), cyanide (CN^-), metyrapone (MP), pyridine (Py) and tert-butylisocyanide (t-BuNC) complexes of ferric cytochrome P-450$_{CAM}$, and for cyanide (CN^-), azide (N_3^-), imidazole (Im) and hydroxide (OH^-) complexes of sperm whale and horse myoglobins.
(Source: ref. 91. Reproduced by kind permission of Elsevier Scientific Publishing Co., Netherlands.)

tribution on the heme iron and porphyrin system, presumably through a conformational change of the apo-protein.

Furthermore, it is suggested on the basis of the nature of the hyperfine shift that the electron spin density in the center of the heme is enhanced by the binding of substrate to the enzyme.

The stick diagram shown in Fig. 3.32 represents the positions of hyperfine shifted methyl peaks obtained with various low spin derivatives of cytochrome P-450$_{CAM}$ as well as several derivatives of myoglobin. Clearly, the effects of ligand substitution on the difference in hyperfine shift of the signals due to methyl protons are small in the case of cytochrome P-450$_{CAM}$, indicating that internal axial ligands of the cytochrome coordinate strongly with the heme iron.

As stated above, the 5th ligand of the heme iron of P-450$_{CAM}$ (presumably S^- from cysteine) has the role of maintaining a rich electron spin density in the center of the heme, which is further accelerated by the substrate binding. This might be important for activation of the liganded oxygen molecule on the 6th site in the case of the ferrous oxygenated form.

3.3.7 Magnetic and Molecular Spectroscopic Investigations on P-450 Other than EPR and NMR Studies

Modern physical techniques have been introduced for elucidation of the charge, spin states and molecular multiplicity of P-450. These involve magnetic susceptibility, Mössbauer spectra and magnetic circular dichroism (MCD). Measurements of the magnetic susceptibility and Mössbauer spectra of cytochrome P-450 have been carried out mainly with preparations obtained from *Pseudomonas putida*, since the degree of purification was much better than that with mammalian P-450. From the Mössbauer spectra, Sharrock et al.[104] have reported the following results. Oxidized P-450$_{CAM}$ in the presence of camphor contained a mixture of high spin ($S=5/2$) and low spin ($S=1/2$) ferric heme sites. Removal of the camphor resulted in a conversion from high spin to low spin form. The contents of high spin form and low spin form were temperature-dependent as shown in Table 3.13. Anaerobic reduction of P-450$_{CAM}$ in camphor solution yielded the high spin ferrous ($S=2$) state. Exposure of this preparation to oxygen produced a new complex, the Mössbauer spectra of which were similar to those of oxyhemoglobin. CO-reduced P-450 complex showed similar Mössbauer spectra to those of hemoglobin-CO complex.

TABLE 3.13. Comparison of high spin fractions by calculation and experiment

Temperature (K)	1.5	4.2	15	94	200	253
Calculated high spin fraction (%)	50	51	61	72	74	74
Experimental high spin fraction (%)	45(estimated)	45±7	60	76	77±3	76

(Source: ref. 104. Reproduced by kind permission of the American Chemical Society.)

For investigating the spin states of P-450, especially the reduced form, the magnetic susceptibility is effective. Work of this type has been carried out recently by Champion et al.[105] with P-450$_{CAM}$. Computer analysis using a spin-Hamiltonian gives values of D from the same equation (Eq.(3.6)) as the ferric heme, of between 15 cm^{-1} and 25^{-1}, which agrees with the value of $D=20$ cm^{-1} obtained for reduced P-450$_{CAM}$.

camphor complex by low field variable temperature Mössbauer spectroscopy.

Magnetic circular dichroism is applied to more direct spectral studies or to the differentiation between P-450 and P-448. Dawson et al.[106] have described P-450 and P-448 as spectrally distinct: the mean of the crossover points for P-450 was at 451.5±0.3, whereas that for P-448 was at 450.5±0.6, where the crossover point was defined as the point where MCD_{obs} and CD intersected. Magnetic circular dichroism has the advantage that it is less affected by contaminant cytochrome b_5 and hemoglobin than the difference spectra.

MCD experiments have also been performed by Shimizu et al.[107] on purified P-450, free of other hemoproteins, obtained from phenobarbital-induced rabbit liver microsomes. The MCD spectrum for oxidized P-450 indicated that it contains both ferric low spin and ferrous low spin species. Reduced P-450 is largely in the ferrous high spin state, though a considerable amount of ferric low spin species is also present.

The molar ellipticity per unit magnetic field, $[\theta]_M$, can be expressed as the sum of three terms, known as the A, B and C terms, respectively.[108] The CO-complex of reduced P-450 shows an apparent A term around 450 nm which consists of about 50% C term and 50% of the other terms, indicating that it is not in a purely ferrous low spin state, but contains a high spin species to some extent. This anomaly disappears on conversion to the CO-complex of reduced P-420. Comparison of the MCD spectrum of ethyl isocyanide complex of reduced P-450 with that of P-420 reveals a fundamental difference in their heme environments.

Resonance Raman (RR) spectroscopy, which has been extensively and successfully applied to various hemoproteins by Spiro, Rimai, Kitagawa and others, reveals the vibrational spectra of iron-porphyrin surrounded by an apo-protein. Ozaki et al.[109] have measured the RR spectra of cytochrome $P-450_{CAM}$ and its enzymatically inactive form (P-420) in various oxidation and spin states. They found an anomaly in the "oxidation state marker" of $P-450_{CAM}$ in the ferrous high spin state (1346 cm^{-1}), different from those of other hemoproteins in the same spin state ($\sim$1355 $\sim$1365 cm^{-1}), and correlated this anomaly to the electron-donating ability of the internal ligand S^-.

REFERENCES

Section 3.3
1. T. Omura and R. Sato, *J. Biol. Chem.*, **237**, PC1375 (1962).
2. H. S. Mason, T. Yamano, J. -C. North, Y. Hashimoto and P. Sakagishi, *Oxidases and Related Redox Systems* (ed. T. King, H. S. Mason and M. Morrison), vol. II, p. 879, John Wiley (1965).
3. H. Beinert and R. H. Sands, *Free Radicals in Biological Systems* (ed. M. S. Blois, Jr., H. W. Brown, R. M. Lemmon, R. Q. Lindblom and M. Weissbluth), p. 17, Academic Press (1961).
4. A. Ehrenberg, *Arkiv Kemi*, **19**, 119 (1962).
5. M. Kotani, *Advan. Chem. Phys.*, **7**, 159 (1964); *Progr. Theoret. Phys. (Kyoto)*, Suppl. 17, 4 (1964).
6. J. S. Griffith, *Nature*, **180**, 30 (1957).
7. Y. Hashimoto, T. Yamano and H. S. Mason, *J. Biol. Chem.*, **237**, PC3843 (1962).
8. R. Bois-Poltoratsky and A. Ehrenberg, *Eur. J. Biochem.*, **2**, 361 (1967).
9. Y. Ichikawa and T. Yamano, *Arch. Biochem. Biophys.*, **121**, 742 (1967).
10. F. Wada, T. Higashi, Y. Ichikawa, K. Tada and Y. Sakamoto, *Biochim. Biophys. Acta*, **88** 654 (1964).
11. R. Tsai, C. -A. Yu, I. C. Gunsalus, J. Peisach, W. Blumberg, W. H. Orme-Johnson and H. Beinert, *Proc. Natl. Acad. Sci. U.S.A.*, **66**, 1157 (1970).
12. M. R. Waterman and H. S. Mason, *Biochem. Biophys. Res. Commun.*, **39**, 450 (1970).
13. Y. Ichikawa and T. Yamano, *J. Biochem. (Tokyo)*, **71**, 1053 (1972).
14. J. Peisach and W. E. Blumberg, *Proc. Natl. Acad. Sci. U.S.A.*, **67**, 172 (1970).
15. M. Kotani, *Advan. Quantum Chem.*, **4**, 227 (1968).
16. T. Iizuka, *Seikagaku* (Japanese), **47**, 1061 (1975).
17. L. Pauling and C. D. Coryell, *Proc. Natl. Acad. Sci. U.S.A.*, **22**, 210 (1936).
18. H. Taube, *Chem. Rev.*, **50**, 69 (1952).
19. J. S. Griffith, *J. Inorg. Nucl. Chem.*, **2**, 1 (1965).
20. J. S. Griffith, *Proc. Roy. Soc., Ser. A*, **235**, 23 (1956).
21. P. George, J. Beetlestone and J. S. Griffith, *Haematin Enzymes* (ed. J. E. Falk, R. Lemberg and R. K. Morton), p. 105, Pergamon Press (1961).
22. J. Beetlestone and P. George, *Biochemistry*, **3**, 707 (1964).
23. D. J. E. Ingram and J. E. Bennett, *Discussions Faraday Soc.*, **19**, 140 (1955).
24. J. Gibson, D. J. E. Ingram and D. Schonland, *ibid.*, **26**, 72 (1958).
25. G. Lang and W. Marshall, *Proc. Phys. Soc.* (London), **87**, 3 (1966).
26. H. Theorell and A. Ehrenberg, *Acta Chem. Scand.*, **5**, 823 (1951).
27. W. Scheler, G. Schoffa and F. Jung, *Biochem. Z.*, **329**, 232 (1957).
28. A. S. Brill and R. J. P. Williams, *Biochem. J.*, **78**, 246 (1961).
29. D. J. E. Ingram and J. C. Kendrew, *Nature*, **178**, 905 (1956).
30. T. Yonetani and H. Schleyer, *J. Biol. Chem.*, **242**, 3926 (1967).
31. T. Yonetani, H. Schleyer, B. Chance and A. Ehrenberg, *Hemes and Hemoproteins* (ed. B. Chance, R. W. Estabrook and T. Yonetani), p. 293, Academic Press (1966).
32. T. Yonetani, H. Schleyer and A. Ehrenberg, *J. Biol. Chem.*, **241**, 3240 (1966).
33. T. Yonetani, D. F. Wilson and B. Seemonds, *ibid.*, **241**, 5347 (1966).
34. T. Iizuka and M. Kotani, *Biochim. Biophys. Acta*, **154**, 417 (1968).
35. T. Iizuka and M. Kotani, *ibid.*, **181**, 275 (1969).
36. T. Iizuka, M. Kotani and T. Yonetani, *ibid.*, **167**, 257 (1968).
37. T. Iizuka and M. Kotani, *ibid.*, **194**, 351 (1969).

38. M. Kotani, *Rev. Mod. Phys.*, **35**, 717 (1963).
39. H. Morimoto, T. Iizuka, J. Otsuka and M. Kotani, *Biochim. Biophys. Acta*, **102**, 624 (1965).
40. H. Glaumann, B. Kuylenstierna and G. Dallner, *Life Sci.* (*II*), **8**, 1309 (1969).
41. Y. Ichikawa, in preparation.
42. A. P. Poland, E. E. Glover, J. R. Robinson and D. W. Nebert, *J. Biol. Chem.*, **249**, 5599 (1974).
43. D. W. Nebert and H. Kon, *ibid.*, **248**, 169 (1973).
44. D. W. Nebert, J. R. Robinson and H. Kon, *ibid.*, **248**, 7637 (1973).
45. J. Peisach, C. A. Appleby and W. E. Blumberg, *Arch. Biochem. Biophys.*, **150**, 725 (1972).
46. K. Murakami and H. S. Mason, *J. Biol. Chem.*, **242**, 1102 (1967).
47. A. Hayashi, A. Shimizu, Y. Yamamura and H. Watari, *Biochim. Biophys. Acta*, **102**, 626 (1965).
48. Y. Ichikawa and J. S. Loehr, *Biochem. Biophys. Res. Commun.*, **46**, 1187 (1972).
49. A. Isomoto, H. Watari and K. Kotani, *J. Phys. Soc. Japan*, **29**, 1571 (1970).
50. T. C. Hollocher, *J. Biol. Chem.*, **241**, 1958 (1966).
51. Y. Ichikawa and T. Yamano, *Biochim. Biophys. Acta*, **131**, 490 (1967).
52. E. Bayer, H. A. O. Hill, A. Roeder and R. J. P. Williams, *Chem. Commun.*, **1969**, 109.
53. C. R. E. Jefcoate and J. L. Gaylor, *Biochemistry*, **8**, 3464 (1969).
54. T. Yamano, *Oxidases and Related Redox Systems* (ed. T. King, H. S. Mason and M. Morrison), vol. 1, p. 262, University Park Press (1971).
55. E. Bayer, P. Krauss, A. Roeder and P. Schretzmann, *ibid.*, p. 227 (1971).
56. J. O. Stern and J. Peisach, *J. Biol. Chem.*, **249**, 7495 (1974).
57. J. P. Collman and T. N. Sorrell, *J. Am. Chem. Soc.*, **97**, 913 (1975).
58. J. P. Collman and T. N. Sorrell, *ibid.*, **97**, 4133 (1975).
59. Y. Ichikawa, B. Hagihara, K. Mori and T. Yamano, *Biochemical and Chemical Aspects of Oxygenases* (ed. K. Bloch and O. Hayaishi), p. 211, Maruzen (1966).
60. K. Dus, M. Katagiri, C. -A. Yu, D. L. Erbes and I. C. Gunsalus, *Biochem. Biophys. Res. Commun.*, **40**, 1423 (1970).
61. H. Schleyer, D. Y. Cooper and O. Rosenthal, *Ann. N.Y. Acad. Sci.*, **222**, 102 (1973).
62. I. C. Gunsalus, I. R. Mecks, J. D. Lipscomb, P. Debrunner and E. Munck, *Molecular Mechanisms of Oxygen Activation* (ed. O. Hayaishi), p. 559, Academic Press (1974).
63. Y. Ichikawa and T. Yamano, *Biochim. Biophys. Acta*, **153**, 753 (1968).
64. L. M. Reichman, B. Annaev and E. G. Rozantzev, *ibid.*, **263**, 41 (1972).
65. K. Comai and J. L. Gaylor, *J. Biol. Chem.*, **248**, 4947 (1973).
66. Y. Miyake, J. L. Gaylor and H. S. Mason, *ibid.*, **243**, 5788 (1968).
67. R. E. Ebel, D. H. O'Keeffe and J. A. Peterson, *FEBS Lett.*, **55**, 198 (1975).
68. H. Kon, *J. Biol. Chem.*, **243**, 4350 (1968).
69. T. Shiga, K. -J. Hwang and I. Tyuma, *Biochimstry*, **8**, 378 (1969).
70. J. Peisach, J. O. Stern and W. E. Blumberg, *Drug Metab. Disposition*, **1**, 45 (1973).
71. R. Tsai, C. -A. Yu, I. C. Gunsalus, J. Peisach, W. Blumberg, W. H. Orme-Johnson and H. Beinert, *Proc. Natl. Acad. Sci. U.S.A.*, **66**, 1157 (1970).
72. S. B. Oldham, L. D. Wilson, W. L. Landgraf and B. W. Harding, *Arch. Biochem. Biophys.*, **123**, 484 (1968).
73. R. W. Estabrook, T. Matsubara, J. I. Mason, J. Werringloer and J. Baron, *Drug Metab. Disposition*, **1**, 98 (1973).
74. R. W. Estabrook, *Oxidases and Related Redox Systems* (ed. T. King, H. S. Mason and M. Morrison), vol. II, p. 511, University Park Press (1973).
75. M. R. Waterman, V. Ullrich and R. W. Estabrook, *Arch. Biochem. Biophys.*, **155**, 355 (1973).
76. J. O. Stern, J. Peisach, W. E. Blumberg, A. Y. H. Lu and W. Levin, *ibid.*, **156**, 404 (1973).

77. D. Ryan, A. Y. H. Lu, J. Kawalek, S. B. West and W. Levin, *Biochem. Biophys. Res. Commun.*, **64**, 1134 (1975).
78. C. Hashimoto and Y. Imai, *ibid.*, **68**, 821 (1976).
79. D. A. Haugen and M. J. Coon, *J. Biol. Chem.*, **251**, 7929 (1976).
80. W. L. Dean and M. J. Coon, *ibid.*, **252**, 3255 (1977).
81. C. R. Jefcoate, E. R. Simpson, G. S. Boyd, A. C. Brownie and W. H. Orme-Johnson, *Ann. N. Y. Acad. Sci.*, **212**, 243 (1973).
82. J. J. Bell, S. C. Cheng and B. W. Harding, *ibid.*, **212**, 290 (1973).
83. F. Mitani, N. Ando and S. Horie, *ibid.*, **212**, 208 (1973).
84. A. C. Brownie, J. Altano, C. R. Jefcoate, W. H. Orme-Johnson and H Beinert, *ibid.*, **212**, 344 (1973).
85. C. R. Jefcoate, W. H. Orme-Johnson and H. Beinert, *J, Biol. Chem.*, **251**, 3706 (1976).
86. D. P. Paul, S. Gallant, N. R. Orme-Johnson, W. H. Orme-Johnson and A. C. Brownie, *ibid.*, **251**, 7120 (1976).
87. M. Katagiri, S. Takemori, E. Itagaki, K. Suhara, T. Gomi and H. Sato, *Iron and Copper Proteins, Advances in Experimental Medicine and Biology* (ed. K. Yasunobu, H. F. Mower and O. Hayaishi), vol. 74, p. 281, Plenus Press (1976).
88. T. Sugiyama, R. Miura and T. Yamano, *ibid.*, p. 290, Plenum Press (1976).
89. S. C. Chang and B. W. Harding, *J. Biol. Chem.*, **248**, 7263 (1973).
90. V. Ullrich, P. Weber and P. Wollenberg, *Biochem. Biophys. Res. Commun.*, **64**, 808 (1975).
91. Y. Ichikawa and T. Yamano, *Biochim. Biophys. Acta*, **131**, 490 (1967).
92. K. Wüthrich, *Struct. Bonding*, **8**, 53 (1970).
93. S. Ogawa and R. G. Shulman, *J. Mol. Biol.*, **70**, 315 (1972).
94. R. G. Shulman, K. Wüthrich, T. Yamane, E. Antonini and M. Brunori, *Proc. Natl. Acad. Sci. U.S.A.*, **63**, 623 (1969).
95. C. C. McDonald and W. D. Phillips, *Biochemistry*, **12**, 3170 (1973).
96. R. M. Keller and K. Wüthrich, *Biochim. Biophys. Acta*, **285**, 326 (1972).
97. R. M. Keller, O. Groudinsky and K. Wüthrich, *ibid.*, **328**, 233 (1973).
98. I. Morishima and T. Iizuka, *J. Am. Chem. Soc.*, **96**, 5279 (1974).
99. T. Iizuka and I. Morishima, *Biophim. Biophys. Acta*, **371**, 1 (1974).
100. I. Morishima and T. Iizuka, *J. Am. Chem. Soc.* **96**, 7365 (1974).
101. I. Morishima and T. Iizuka, *Biochim. Biophys. Acta*, **386**, 542 (1975).
102. I. Morishima, T. Iizuka and Y. Ishimura, *Seikagaku* (Japanese), **46**, 535 (1974).
103. R. G. Shulman, S. H. Glarum and M. Karplus, *J. Mol. Biol.*, **57**, 93 (1971).
104. M. Sharrock, E. Munck, P. G. Debrunner, V. Marshall, J. D. Lipscomb and I. C. Gunsalus, *Biochemistry*, **12**, 258 (1973).
105. P. M. Champion, E. Munck, P. G. Debrunner, T. H. Moss, J. D. Lipscomb and I. C. Gunsalus, *Biochim. Biophys. Acta*, **376**, 579 (1975).
106. J.H. Dawson, P.M. Dolinger, J.R. Trudell, G. Barth, R.E.Linder, E. Bunnenberg and C. Djerassi, *Proc. Natl. Acad. Sci U.S.A.*, **71**, 4594 (1974).
107. T. Shimizu, T. Nozawa, M. Hatano, Y. Imai and R. Sato, *Biochemistry*, **14**, 4172 (1975).
108. C. Djerassi, E. Bunnenberg and D. L. Elder, *Pure. Appl. Chem.*, **25**, 57 (1971).
109. Y. Ozaki, T. Kitagawa, Y. Kyogoku, H. Shimada. T. Iizuka and Y. Ishimura, *J. Biochem. (Tokyo)*, **80**, 1447 (1976).

CHAPTER 4

Cytochrome P-450-containing Oxygenase Systems

4.1. Hepatic Microsomal Systems*[1]
  4.1.1. Electron Transport System of Liver Microsomes
  4.1.2. Electron Transfer from NAD(P)H to Cytochrome P-450 in Liver Microsomes
  4.1.3. Reconstitution of Oxygenase Activity from Solubilized Microsomal Components
  4.1.4. Substrate Specificity of the Hepatic Microsomal Cytochrome P-450 System
4.2. Adrenocortical Mitochondrial Systems*[2-4]
  4.2.1. Sequential Fragmentation of Steroid Hydroxylating Components of Adrenal Cortex
  4.2.2. Adrenodoxin
  4.2.3. NADPH-Adrenodoxin Reductase
  4.2.4. Cytochrome P-450
  4.2.5. General Discussion
4.3. Bacterial Systems*[4]
  4.3.1. Catalytic Components of the Camphor-Methylene Hydroxlating System (Camphor 5-exo Hydroxylase)
  4.3.2. Function of $P\text{-}450_{CAM}$
  4.3.3. Cytochrome P-450-dependent n-Alkane Methyl Hydroxylation (ω-Hydroxylation)
  4.3.4. Other Bacterial P-450-type Cytochromes
  4.3.5. Types of Bacterial Cytochrome P-450-containing Hydroxylase Systems
4.4. Yeast Microsomal Systems*[5]
  4.4.1. General Considerations on Yeast Microsomal Electron Transport Systems
  4.4.2. Purification and Properties of Yeast Cytochrome P-450
  4.4.3. Cytochrome P-450-containing Electron Transport Systems of Yeasts
  4.4.4. Function of Cytochrome P-450-containing Systems of Yeasts
4.5. Induction and Disappearance of Cytochrome P-450 in Yeast Cells*[6]
  4.5.1. Induction of Cytochrome P-450 in Alkane-utilizing Yeast
  4.5.2. Changes in the Cytochrome P-450 Content of Facultatively Aerobic Yeast Cells Grown under Various Conditions
  4.5.3. Physiological Significance of Changes in Cytochrome P-450 Content

4.1 Hepatic Microsomal Systems

4.1.1 Electron Transport System of Liver Microsomes

Soon after establishing the cell fractionation procedure for animal tissues, Hogeboom and his associates discovered the presence of a strong NADH-cytochrome c reductase activity[1] and an NADPH-cytochrome c reductase activity[2] in "submicroscopic particles" prepared from liver homogenates by fractional centrifugation. These particles, which we now call microsomes, were identified by Palade and Siekevitz[3] as fragmented vesicles of endoplasmic reticulum. Since cytochrome c was exclusively found in the mitochondrial fraction,[4] and cytochrome c oxidase activity was also associated with mitochondria,[5,6] the physiological significance of these microsomal cytochrome c reductase activities remained unclear. The microsomal reductase activities were insensitive to Antimycin A,[7] so that the mechanism of electron transfer from NADH or NADPH to cytochrome c catalyzed by microsomes was different from the corresponding mitochondrial reductase activities.

The first important step in the elucidation of the molecular constitution of the microsomal electron transport system was made by C.F. Strittmatter and Ball[8] in 1952, when they discovered the presence of a unique b-type cytochrome in the microsomal fraction prepared from rat liver. This cytochrome, which we now call cytochrome b_5, was different from mitochondrial cytochrome b. It had been previously noticed by Yoshikawa[9] in 1951 on the basis of spectroscopic examinations of slices and Keilin-Hartree preparations of liver and kidney. The discovery of this unique cytochrome in liver microsomes was quickly followed and expanded by P. Strittmatter and Velick who solubilized and purified it.[10] They also isolated from microsomes a new flavoprotein which

[*1] Tsuneo OMURA, Department of Biochemistry, Faculty of Science, Kyushu University, Hakozaki, Fukuoka 812, Japan
[*2] Shigeki TAKEMORI, Department of Environmental Sciences, Faculty of Integrated Arts and Sciences, Hiroshima University, Higashisenda-cho, Hiroshima 730, Japan
[*3] Katsuko SUHARA, Department of Chemistry, Faculty of Science, Kanazawa University, Kanazawa 920, Japan
[*4] Masayuki KATAGIRI, Department of Chemistry, Faculty of Science, Kanazawa University, Kanazawa 920, Japan
[*5] Yuzo YOSHIDA, Faculty of Pharmacy, Mukogawa Women's University, Nishinomiya 663, Japan
[*6] Kunio TAGAWA, 2nd Department of Biochemistry, Osaka University Medical School, Kita-ku, Osaka 530, Japan

effectively reduced cytochrome b_5 by NADH.[11] Since the reduced form of this cytochrome reacted very rapidly with externally added cytochrome c, the presence of NADH-cytochrome b_5 reductase and cytochrome b_5 in microsomes explained the strong NADH-cytochrome c reductase activity of liver microsomes.[11]

On the other hand, the NADPH-cytochrome c reductase activity of liver microsomes was the activity of a single enzyme. Isolation of this reductase from liver microsomes was first reported in 1962 by Williams and Kamin[12] and also by Phillips and Langdon.[13] The properties of the purified enzyme suggested its identity with a flavoprotein which had previously been isolated from acetone powder of whole liver by Horecker.[14] The activity of the isolated NADPH-cytochrome c reductase was sufficiently strong to explain the microsomal reduction of externally added cytochrome c by NADPH. The reductase was also active in reducing various artificial electron acceptors such as ferricyanide, quinones, and dyes. In the presence of catalytic amounts of certain naphthoquinones including vitamin K_3, this flavoprotein could also catalyze rapid oxidation of NADPH by molecular oxygen.[15] However, as in the case of the NADH-cytochrome c reductase system, the absence of endogenous cytochrome c in microsomes placed doubt on the physiological significance of this reductase in them, and the presence of these electron transport enzymes in liver microsomes attracted little attention among biochemists.

The discovery of another new cytochrome, cytochrome P-450, in liver microsomes[16,17] and the elucidation of its participation in NADPH-linked oxygenation reactions,[18,19] opened a new area in the study of microsomal electron transport systems. The presence of this unique carbon monoxide-binding pigment in liver microsomes was first described by Klingenberg[20] and also by Garfinkel[21] in 1958, although its unusual spectral properties and instability prevented elucidation of its hemoprotein nature at the time. The presence of this unusual hemoprotein in liver microsomes was also independently discovered from its characteristic ESR signals by Mason and his associates,[22] who tentatively named it "microsomal Fe_x". The identity of microsomal Fe_x with cytochrome P-450 was later established from several lines of evidence[23,24].

After the discovery of its physiological function as the terminal oxidase of the microsomal NADPH-dependent oxygenase system, cytochrome P-450 attracted the attention of many biochemists, and the electron transport system in liver microsomes also became the subject of intensive studies. Soon the role of NADPH-cytochrome c reductase in the transfer of electrons from NADPH to cytochrome P-450 was established from various lines of evidence, as described in detail below in section 4.1.2. The main physiological function of this microsomal flavoprotein was not the reduction of cytochrome c, but the supply of electrons from NADPH to cytochrome P-450 for oxygenation reactions. Thus cytochrome P-450 and NADPH-cytochrome c reductase constitute the microsomal NADPH-linked oxygenase system, for which an alternate name, NADPH-cytochrome P-450 reductase, sounds more appropriate. However, reduction of cytochrome P-450 does not seem to be the sole function of this reductase. Cytochrome b_5 in microsomes receives electrons from NADPH via this reductase,[25,26] and NADPH-linked peroxidation of microsomal lipids is also dependent on it.[27] The involvement of this enzyme in microsomal squalene epoxidase[28] and heme oxygenase[29] reactions has recently been shown. Apparently, this reductase is functional in supplying electrons to various terminal electron acceptors in microsomes which require the provision of reducing equivalents by NADPH.

Studies on the physiological function of the microsomal NADH-linked electron transport system have also made considerable progress. Although liver microsomes catalyze very rapid reduction of externally added cytochrome c by NADH, this activity cannot be the function of physiological importance. Other electron acceptors for this reductase system have been sought. Since cytochrome b_5 in perfused rat liver remained fully reduced even when the oxygen tension in the circulating blood was increased to a high level[30], active participation of this cytochrome in cellular respiration of hepatocytes was discounted. The first candidate proposed as a physiological electron acceptor for this reductase system was stearyl coenzyme A desaturase[31,32] which was also called "cyanide-sensitive factor"[33] This proposal has been substantiated by later studies,[34-36] and a few more microsomal metabolic reactions have so far been suggested to receive electrons from NADH via NADH-cytochrome b_5 reductase and cytochrome b_5[37] The possible role of cytochrome b_5 in the supply of electrons from NADH to cytochrome P-450 will be discussed in section 4.1.2.

Thus, the main physiological role of cytochrome b_5 and its reductase in microsomes is also to supply reducing equivalents from NADH to various metabolic reactions catalyzed by microsome-bound enzymes. Such a function of the microsomal NADH-cytochrome c reductase system was already predicted by Strittmatter and Ball[38] in 1954. They stated in one of their publications on microsomal cytochrome,[38] "Electrons available from DPNH may not be funneled through cytochrome c to oxygen but may instead be utilized for purposes of reductive synthesis", although no solid evidence for such a concept was available at the time. Their hypothesis has now been fully substantiated with additional supporting evidence.

On the basis of these experimental data accumulated in the past

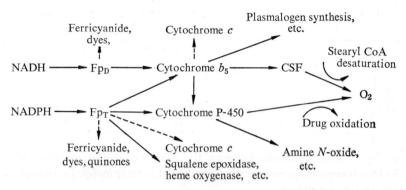

Fig. 4.1. Sequence of electron transfer in the microsomal electron transport system. Fp_D=NADH-cytochrome b_5 reductase; Fp_T=NADPH-cytochrome c reductase; CSF=cyanide-sensitive factor (stearyl CoA desaturase). The dashed lines indicate electron flow to non-physiological electron acceptors.

decade, it is now possible to present a scheme for the pathways of electron flow in the microsomal electron transport system, as shown in Fig. 4.1. The abbreviations Fp_D and Fp_T for NADH- and NADPH-specific flavoproteins, respectively, are taken from a review by Siekevitz.[39] Various lines of indirect evidence frequently suggested the presence of an unknown factor "X" between Fp_T and cytochrome P-450.[27,40] However, recent studies with purified Fp_T and cytochrome P-450 now indicate their direct interaction in the transfer of electrons from NADPH to cytochrome P-450 as will be described in section 4.1.2.

All electrons for the microsomal electron transport system are derived from NADH or NADPH through either of these flavoproteins (Fig. 4.1). NADH→Fp_D→cytochrome b_5 and NADPH→Fp_T→cytochrome P-450 form two major paths for electrons in the system, and the electrons spread out from these two major paths like an unfolded fan to multiple terminal electron acceptors to be utilized in various metabolic reactions which require the supply of reducing equivalents. Since the endoplasmic reticulum in the liver cell is surrounded by cytoplasm where active generation of the reduced forms of nicotinamide adenine dinucleotides is occurring, the microsomal electron transport system is always provided with an ample supply of reducing equivalents from the cytoplasm.

The electron flow in the membrane of liver microsomes is thus in sharp contrast to the mitochondria-type electron transport chain where electrons flow into the chain at various levels through substrate-linked dehydrogenases and are then funneled through cytochrome c into a single terminal electron acceptor, cytochrome c oxidase. Such a difference

TABLE 4.1. Contents of cytochromes P-450 and b_5, NADPH-cytochrome c and NADH-cytochrome b_5 reductases in rat liver microsomes

Enzymes	Content in microsomes	
	(nmole/mg protein)	(% of total protein)
Cytochrome P-450	1.0	5.0
NADPH-cytochrome c reductase	0.05	0.5
Cytochrome b_5	0.5	1.0
NADH-cytochrome b_5 reductase	0.05	0.2

in physiological functions between the microsomal and mitochondrial electron transport systems should be reflected in their molecular organization in the membrane.

Table 4.1 shows the contents of cytochrome P-450, cytochrome b_5, NADPH-cytochrome c reductase, and NADH-cytochrome b_5 reductase in rat liver microsomes. Since these enzymes have already been purified and their molecular weights are known, it is possible to calculate their contents in microsomes on a molar basis as well as on a weight basis. It is immediately clear from the table that the molar ratios of the cytochromes to the correponding flavoproteins are much larger than unity: 10-20 mole cytochrome P-450 or cytochrome b_5 are present per mole of NADPH- or NADH-linked reductase, respectively. Moreover, the molar ratios of these metabolically linked pairs of enzymes vary considerably according to variations in the physiological condition of the animals. Since the combined amount of these enzymes, which constitute the main part of the microsomal electron transport system, is less than 10% of the total membrane protein, they are unable to cover the whole surface of the endoplasmic reticulum. This situation raises an interesting question concerning the reactions among these enzymes in the membranes of microsomes.

A brief survey of present knowledge on the molecular architecture of microsomal membranes will first be given. In general, the membrane model proposed by Singer[41] is applicable to the microsomal membrane. Cytochrome b_5, NADH-cytochrome b_5 reductase, and NADPH-cytochrome c reductase are typical "integral" membrane proteins in that their molecules are amphipatic and attach to the membrane by hydrophobic interaction of the hydrophobic parts of the molecules with the membrane matrix. Actually, elucidation of the molecular properties of native cytochrome b_5 by Ito and Sato[42] was one of the earliest pieces of evidence for the amphipatic nature of membrane proteins and contributed to the formulation of Singer's model of biological membranes.

As seen from Table 4.2, each of the molecules of these microsomal enzymes consists of a major hydrophilic portion, in which the catalyti-

TABLE 4.2. Amphipatic nature of molecules of cytochrome b_5, NADPH-cytochrome c and NADH-cytochrome b_5 reductases

Enzyme	Molecular weight		Ref.
	Native form	Hydrophilic part (protease-solubilized)	
Cytochrome P-450	50,000	—	43
Cytochrome b_5	17,000	12,000	44
NADPH-cytochrome c reductase	80,000	71,000	45
NADH-cytochrome b_5 reductase	43,000	33,000	46

cally active part resides, and a minor hydrophobic portion which is essential for attachment of the molecule to the membrane.[47,48] The hydrophilic portion of the enzyme molecule can be selectively split from its hydrophobic portion by digestion with a suitable protease (Table 4.2), resulting in solubilization of the active fragment of the enzyme from the microsomes. Since the microsomal vesicles were kept closed with morphologically well-defined membranes even after proteolytic digestion,[49] the efficient solubilization of these microsomal enzymes on protease treatment was taken as evidence for their external localization in the microsomal vesicles.[50] The major hydrophilic portions, the catalytically active portions, of all these enzymes must therefore be protruding from the membrane into the surrounding aqueous environment of the membrane vesicles, which is the cytoplasm of the cell. The external localization of these microsomal enzymes has also been confirmed by the effective inhibition of their catalytic activities in microsomes with antibodies prepared against the hydrophilic portions of the enzyme molecules.[36,51,52] The molecule of cytochrome P-450 is highly hydrophobic. However, since this cytochrome receives electrons from NADPH-cytochrome c reductase in the same membrane vesicle, some portion of the cytochrome P-450 molecule must also be protruding from the membrane into the surrounding aqueous environment.

Electron transfer reactions among these flavoproteins and cytochromes in microsomes require the collision of the catalytically active portion of the membrane-bound enzyme molecules. It is necessary to choose either one of the following two possibilities to explain the electron transfer reactions which occur on the external surface of microsomal vesicles.

(1) The molecules of all components of the microsomal electron transport system are quite randomly distributed in the membrane. Electrons are transferred among them as a result of the random collisions caused by their lateral movements in the membrane.

(2) Molecular aggregates of reacting electron-transferring enzymes exist in the membrane. Electrons are efficiently transferred among the

enzyme molecules forming one aggragate, but not among separate aggregates.

The first possibility emerged from the recent concept of a fluid-like structure for biological membranes.[41] Supporting evidence for this possibility has recently been provided by Rogers and Strittmatter,[53] who carried out *in vitro* addition of large amounts of native cytochrome b_5 to liver microsomes and found no difference between extra bound cytochrome b_5 and the endogenous cytochrome in their kinetics of reduction by NADH. The results indicated that cytochrome b_5 molecules randomly incorporated into the membrane *in vitro* were given an equal opportunity of reaction with NADH-cytochrome b_5 reductase molecules in the membrane to the cytochrome molecules originally present in the microsomes, and are in accord with the concept of a random distribution of these cytochromes and flavoproteins in the microsomal membrane.

However, supporting evidence for the second possibility has also been frequently reported. Subfractionation of microsomes by various means often indicated a non-uniform distribution of cytochromes and flavoproteins among microsomal vesicles. Sucrose density gradient centrifugation of sonicated rat liver microsomes showed a certain degree of separation of NADH-cytochrome b_5 reductase and cytochrome b_5 from NADPH-cytochrome c reductase and cytochrome P-450, suggesting a non-random distribution for these two enzyme groups in the membranes of the endoplasmic reticulum.[54,55] Immunoadsorption of microsomal vesicles to the immobilized antibody against NADH-cytochrome b_5 reductase also indicated a significant tendency for separation of NADH-cytochrome b_5 reductase from NADPH-cytochrome c reductase.[56] These results conflict with a random uniform distribution for the component molecules of the microsomal electron transport system.

Kinetic studies on the reduction of cytochrome P-450 or cytochrome b_5 by NADH or NADPH have also frequently provided evidence for the presence of molecular aggregates of flavoprotein reductases and cytochromes in the membrane. The effect of mersalyl on the electron flow from NADPH to cytochrome P-450[57] indicated the presence of a rigid electron transport complex linking a single molecule of NADPH-cytochrome c reductase to a number of molecules of cytochrome P-450. Selective removal of NADH-cytochrome b_5 reductase from microsomes by cathepsin digestion[58] suggested the presence of an assembly of NADH-cytochrome b_5 reductase and cytochrome b_5 in which several molecules of the reductase were associated with about ten times as many molecules of cytochrome b_5. A spin label study on microsomal phospholipid[59] indicated that the molecule of NADPH-cytochrome c reductase in the microsomal membrane is surrounded by an unusual rigid quasicrystalline halo of phospholipid at a temperature where the bulk membrane lipid is in a fluid state. This observation is interpreted to indicate the surrounding of the reductase molecule with several molecules of cytochrome P-450.

However, none of the experimental evidence presented so far has been sufficiently conclusive to settle this problem. More detailed information on the distribution of flavoproteins and cytochromes in microsomal membranes is surely needed, and the actual situation may in fact lie between the extreme models visualized by the first and second posibilities. The molecules of the component enzymes of the microsomal electron transport system may perhaps be in a dynamic state of association and dissociation in the membrane. The specific protein-protein interactions between interacting enzyme molecules favor the association of NADH-cytochrome b_5 reductase with cytochrome b_5 or NADPH-cytochrome c reductase with cytochrome P-450, leading to the formation of multimolecular electron transfer complexes of varying sizes in the membrane, whereas the freedom of lateral movement of membrane-bound enzyme molecules tends to dissociate these molecular aggregates. Such a concept of a loose dynamic assembly of enzyme molecules in the membrane may serve to reconcile the conflicting reports for and against the presence of multi-molecular electron transport complexes in microsomes.

4.1.2 Electron Transfer from NAD(P)H to Cytochrome P-450 in Liver Microsomes

In the earliest papers reporting the presence of cytochrome P-450 in liver microsomes, the reducibility of this pigment by NADH[20,21] and by NADPH[16] was described. Since it was difficult to observe the reduced minus oxidized difference spectrum of cytochrome P-450 in microsomes, the reduction of the cytochrome could be detected only in the presence of carbon monoxide by measuring the appearance of the 450 nm absorption peak of the carbon monoxide compound of the reduced pigment. The oxidation of the reduced cytochrome P-450 by molecular oxygen was also determined by measuring the disappearance of the absorption at 450 nm of the carbon monoxide compound.[16] The affinity of reduced cytochrome P-450 for carbon monoxide was high,[60] while the oxidized form of the cytochrome did not combine with carbon monoxide. The combination of reduced cytochrome P-450 with carbon monoxide was sufficiently rapid[60] to allow the formation of the carbon monoxide compound to be utilized in determining the rate of reduction of the cytochrome in microsomes by NADH or by NADPH.

The reduction of microsome-bound cytochrome P-450 by NADPH showed complicated reaction kinetics. The time course of the reduction was not a simple monophasic curve agreeing with first order reaction kinetics,[61] and indicated the presence of at least two fractions of cytochrome P-450 with widely different rates of reduction.[61] This biphasicity was noticed even when sodium dithionate was used as the reducing agent,[62] and when reduction of the cytochrome P-450 was directly monitored by measuring the optical density change at 415 nm in the absence of carbon monoxide.[62]

The rate of reduction of cytochrome P-450 in microsomes could be modified by the presence of certain compounds which combine with the oxidized form of the cytochrome. Various substrates (ethylmorphine, hexobarbital, SKF-525A, etc.) which induced type I spectral changes of cytochrome P-450, increased the rate of its reduction by NADPH, whereas some substrates (aniline, nicotinamide, DPEA, etc.) which were capable of inducing type II spectral changes upon combination with cytochrome P-450, decreased the rate of reduction of the cytochrome.[61] In the presence of ethyl isocyanide, which also combined with cytochrome P-450, the reduction of the cytochrome was stimulated and the time course of reduction showed a monophasic curve which obeyed first order reaction kinetics.[62]

The physiological significance of NADPH-linked reduction of microsomal cytochrome P-450 was demonstrated by the discovery of the function of this hemoprotein in the NADPH-dependent oxygenation of steroids and various drugs.[18,19] Since NADPH-cytochrome c reductase was the only known microsomal enzyme which could be regarded as functional in the transfer of electrons from NADPH to cytochrome P-450, the common view was for the participation of this flavoprotein in the reduction of cytochrome P-450 in microsomes. This inference received support from various lines of evidence in the subsequent years.

The first, indirect evidence was the inhibition of the NADPH-dependent drug oxidation activities of microsomes by ferri-cytochrome c.[13,63,64] Since cytochrome c could efficiently take electrons away from reduced NADPH-cytochrome c reductase, Phillips and Langdon[13] ascribed the inhibition of microsomal hydroxylation reactions by cytochrome c to a competition between endogenous hydroxylase (cytochrome P-450) and externally added cytochrome c for the supply of electrons from this flavoprotein.

The next evidence for the role of NADPH-cytochrome c reductase in the reduction of cytochrome P-450 came from the finding of the selective removal of this flavoprotein from microsomes by relatively mild protease treatment. NADPH-cytochrome c reductase could be easily solubilized from microsomes by digestion with low concentrations of Steapsin.[65] Although cytochrome P-450 was neither solubilized nor significantly converted to P-420 by the digestion, its rate of reduction by NADPH was lowered in proportion to the loss of NADPH-cytochrome c reductase from the microsomes.[66] Concomitantly, the digested microsomes lost their activity for NADPH-dependent hydroxylation of aniline.[66] This observation was later confirmed using trypsin,[67] which was also effective in selectively solubilizing NADPH-cytochrome c reductase from microsomes.[49]

The third evidence, which was also indirect, was presented by Ernster and Orrenius[27] on the basis of a study of the induction of the microsomal oxygenase system by phenobarbital administration to rats. The parallel increase in cytochrome P-450 and NADPH-cytochrome c reductase in the drug-treated animals was regarded as indicating the participation of these two microsomal components in NADPH-dependent drug oxidation reactions.

Soon, two separate and new approaches i.e. an immunochemical study on microsomal flavoproteins and a reconstitution experiment on the microsomal oxygenase system from solubilized components, yielded decisive evidence for the participation of NADPH-cytochrome c reductase in the electron transfer from NADPH to cytochrome P-450 in the microsomal system. An antiserum was prepared against trypsin-solubilized NADPH-cytochrome c reductase, and was found to inhibit the cytochrome c reductase activity of the membrane-bound reductase as well as that of the solubilized enzyme employed as the antigen.[51] This antibody was used in examining the participation of the reductase in the NADPH-dependent hydroxylation of aniline by liver microsomes,[68] and the inhibition of the hydroxylation reaction by the antibody strongly suggested the participation of the reductase. These findings were later confirmed and extended to other NADPH-dependent microsomal oxygenase activities,[69-71] although a non-inhibitory antibody has been described in one case.[72]

The solubilization and fractionation of the microsomal oxygenase system into its individual components, and the reconstitution of the active enzyme complex from the isolated components were first carried out successfully in 1968 by Lu and Coon.[73] They identified one of the essential components as NADPH-cytochrome c reductase, confirming the participation of this flavoprotein in the transfer of electrons from NADPH to cytochrome P-450 for oxygenation reactions. Their reconstitution experiments will be described in detail in section 4.1.3.

When these studies aiming at characterization of the NADPH-linked reducing system for microsomal cytochrome P-450 were being carried out in the latter half of the 1960's, a soluble NADPH-cytochrome P-450 reductase system was isolated from the mitochondria of bovine adrenal cortex.[74] This reductase system consisted of two components, a flavoprotein with NADPH-diaphorase activity and a ferredoxin-type iron-sulfur protein, and was active in the reduction of mitochondrial membrane-bound cytochrome P-450 by NADPH (cf. section 4.2.). Two years later, a soluble camphor oxygenase system containing cytochrome P-450 was isolated from a bacterium[75] in which the electrons for the monooxygenation reaction were supplied from NADH to soluble cytochrome P-450 via a flavoprotein and an iron-sulfur protein (cf. section 4.3). These findings aroused renewed interest in the possible presence of an intermediate electron carrier, perhaps an iron-sulfur protein, between NADPH-cytochrome c reductase and cytochrome P-450 in the liver microsomal system. The soluble NADPH-cytochrome P-450 reductase system obtained from bovine adrenal cortex mitochondria could actually reduce the cytochrome P-450 in submicrosomal particles prepared from rat liver.[76]

The presence of such an intermediate electron carrier in the NADPH-linked electron transport pathway of liver microsomes had previously been suggested from an inhibition study of microsomal electron transfer reactions[27] and also from a study of the postnatal development of the microsomal electron transport system in the liver of newborn rats.[40] The hypothetical intermediate electron carrier between NADPH-cytochrome c reductase and cytochrome P-450 was called "X". Since ferredoxin-type iron-sulfur proteins exhibit characteristic ESR signals, detailed examinations of the ESR signals of microsomes were expected to provide a definite answer for or against the presence of such an electron carrier in microsomes. Some investigators reported detection of the expected ESR signals with kidney microsomes,[77] while many others failed to detect any such signal with liver microsomes.[78]

Recent reports on the reconstitution of an NADPH-dependent oxygenase system from purified microsomal components seem to exclude the need for an intermediate electron carrier in the transfer of reducing equivalents from NADPH-cytochrome c reductase to cytochrome P-450, and neither of these two essential components of the oxygenase system contains non-heme iron or acid-labile sulfur in its molecule.[79] Judging from these available data, it may be concluded that the reduction of cytochrome P-450 in liver microsomes by NADPH is probably catalyzed by NADPH-cytochrome c reductase without the participation of any other intermediate electron carriers.

The reduction of microsomal cytochrome P-450 by NADH did

not attract much attention when active studies on the mechanism of reduction of the cytochrome by NADPH were actually in progress. However, the capability of NADH for supporting the microsome-catalyzed oxidation of various drugs was already known in the early 1960's.[64,80,81] NADH alone could sustain the drug oxidation by liver microsomes at 1/2 to 1/5 the level of the corresponding NADPH-dependent activities, and the addition of NADH to the NADPH-containing reaction mixture elevated the rates of drug oxidation to higher levels.[64,81] Since an indirect effect for NADH via NADPH generation by a transhydrogenase reaction was examined and excluded,[64] the presence of a path of electron flow from NADH to cytochrome P-450 in liver microsomes was indicated. These findings coincided with the observed reduction of cytochrome P-450 in isolated microsomes by the addition of NADH.[17]

The first detailed study on this subject was made by Ichikawa et al.[82] who found a very high K_m value for NADH (1.3×10^{-3} M) in the NADH-dependent p-hydroxylation of o-chloroaniline catalyzed by rabbit liver microsomes. The K_m for NADPH was 3.2×10^{-6} M under identical conditions. Since these K_m values for NADH and NADPH coincided with those for isolated NADPH-cytochrome c reductase which was solubilized and purified from the same microsomes,[83] they concluded that the same flavoprotein, NADPH-cytochrome c reductase, is functional in accepting electrons from both NADH and NADPH in order to supply them to cytochrome P-450 for oxygenation reactions. However, in a later publication,[84] Ichikawa and Loehr presented evidence for the participation of NADH-cytochrome b_5 reductase in the reduction of microsome-bound cytochrome P-450 by NADH. They prepared a submicrosomal particle preparation by treating rabbit liver microsomes with proteases in the presence of glycerol,[85] and found that the cytochrome P-450 in the protease-treated particles was still reducible by NADH but not by NADPH. Since the protease treatment removed both NADPH-cytochrome c reductase and cytochrome b_5 from the microsomes almost campletely, they explained the observed NADH-linked reduction of cytochrome P-450 by a direct interaction of the cytochrome with NADH-cytochrome b_5 reductase without the participation of cytochrome b_5.[84]

The second important development in this subject came from the work of Estabrook and his associates in 1971.[86-89] They reported a cooperative interaction between NADH and NADPH in supporting the drug oxidation reactions of liver microsomes, and suggested that the stimulation of NADPH-dependent drug oxidations by NADH was due to a more efficient supply of electrons to the oxygenated form of reduced cytochrome P-450 from NADH than from NADPH. Two reducing equivalents are consumed in the oxidation of one mole of substrate by the microsomal cytochrome P-450 system. According to their scheme (*cf.* chapter 5), the first electron which reduces ferric cytochrome P-450 to its ferrous form comes from NADPH, and the second one which is consumed in the activation of the oxygen molecule bound to cytochrome P-450, can be supplied by either of NADH and NADPH. As the NADPH-supported reduction level of microsomal cytochrome b_5 was lowered by the addition of a suitable substrate for the oxygenase activity of cytochrome P-450, they suggested the need for the participation of cytochrome b_5 in the transfer of the second electron from both NADH and NADPH to cytochrome P-450.[88] Their findings were confirmed by Correia and Mannering,[90,91] who also postulated the participation of cytochrome b_5 in explaining their experimental data, and a new term "NADH synergism" was created to indicate the synergistic effect of NADH on the NADPH-dependent oxidation of drugs by microsomes.

Although these kinetic studies on NADH synergism provided positive evidence for the participation of cytochrome b_5 in the supply of reducing equivalents to cytochrome P-450 for drug oxidation reactions, an opposing view was also expressed[92] based on the finding that the addition of extra cytochrome b_5 to microsomes caused inhibition, not activation, of NADPH-dependent drug oxidations. Staudt *et al.*[93] also proposed a different explanation for NADH synergism in which the observed synergistic action of NADH was ascribed to its electron sparing effect for NADPH by reducing activated oxygen which was not used for monooxygenation reactions. According to this scheme, the synergistic effect of NADH on NADPH-dependent drug oxidation reactions would not be observed with a substrate having a very tight coupling of oxygen activation and substrate hydroxylation since the electrons coming from NADH via cytochrome b_5 are consumed only by the oxygen molecules which have been activated by combination with ferrous cytochrome P-450 but have failed to react with the substrate. To settle this problem, reconstitution and immunochemical experiments again came to the fore.

As described in section 4.1.3, successful reconstitution of the NADH-dependent drug oxidation system was reported by Lu *et al.*[94,95] Their system consisted of NADH-cytochrome b_5 reductase, cytochrome b_5, and cytochrome P-450, and there was no doubt about the capability of NADH-cytochrome b_5 reductase and cytochrome b_5 in transferring electrons from NADH to cytochrome P-450 for drug oxidation reactions. The reconstitution experiments also showed, however, that the inclusion of cytochrome b_5 in the NADPH-dependent drug oxidation system consisting of purified NADPH-cytochrome *c* reductase and cytochrome

P-450, inhibited the NADPH-dependent oxidation of some substrate drugs, and that the inhibition was relieved by adding NADH and NADH-cytochrome b_5 reductase to the reaction mixture.[95] Moreover, NADH synergism could not be observed with the reconstituted system.[95]

Apparently, when we deal with a membrane-bound enzyme system, reconstitution experiments are subject to a certain limitation in that it is usually difficult to establish the formation of the original molecular organization of the enzyme system which is functional in the membrane. Another difficulty in such experiments is the presence of certain amounts of detergents in the purified protein components which may cause inhibitory or stimulatory effects on the interactions among particular components in the reconstituted system. Immunochemical experiments using antibodies which specifically react with the hydrophilic parts of particular membrane-bound enzymes in the intact membrane are useful when the above-mentioned drawbacks of reconstitution experiments render the application of the reconstitution technique difficult.

Immunochemical experiments with antisera against cytochrome b_5[96,97] have provided definite evidence for the participation of this cytochrome in the NADH synergism of cytochrome P-450-catalyzed reactions. The antisera inhibited the NADH-stimulation of NADPH-supported N-demethylation of ethylmorphine, but exerted little effect on the demethylation reaction when NADPH was the only source of reducing equivalents. Undoubtedly, the reducing equivalents from NADH are supplied to cytochrome P-450 via cytochrome b_5 when both NADPH and NADH are present in the reaction mixture. In accordance with the results of the reconstitution studies, the participation of cytochrome b_5 in NADH-dependent oxidation reactions of microsomal cytochrome P-450 was also confirmed by the use of anti-cytochrome b_5 serum.[98] It may thus be concluded that the NADH-cytochrome b_5 reductase and cytochrome b_5 in liver microsomes can be functional in transferring reducing equivalents from NADH to cytochrome P-450, although the role of cytochrome b_5 in the transfer of electrons from NADPH to cytochrome P-450 remains to be elucidated.

4.1.3. Reconstitution of Oxygenase Activity from Solubilized Microsomal Components

Since the membrane-bound nature of microsomal NADPH-dependent oxygenase activity was elucidated, the solubilization of this enzyme activity from microsomes was repeatedly attempted by many investigators. However, most of the conventional solubilization procedures (i.e. cholate treatment, snake venom digestion, organic solvent treatment, etc.) which had been successfully employed in solubilizing particle-bound enzymes from various biological sources resulted in inactivation of this microsomal oxygenase. When the basic properties of cytochrome P-450 and its principal role in the microsomal oxygenase system were elucidated at the beginning of the 1960's, it became evident that the conversion of this cytochrome into its P-420 form was mainly reponsible for the loss of activity of this particle-bound oxygenase system upon solubilization. This difficulty was overcome by the finding of Ichikawa and Yamano[99] in 1967 that cytochrome P-450 could be stabilized by polyols and reduced glutathione against conversion to its P-420 form. They were even able to demonstrate the reconversion of P-420 to the P-450 form on the addition of these reagents to cholate- or PCMB-treated microsomes. Their discovery was quickly utilized in the following year by Lu and Coon[73] in their first successful reconstitution of the cytochrome P-450 oxygenase system from solubilized microsomal components.

These authors[73,100] treated rabbit liver microsomes with sodium deoxycholate in the presence of glycerol and dithiothreitol, and obtained a soluble preparation which catalyzed the NADPH-dependent ω-hydroxylation of lauric acid by molecular oxygen. Moreover, they succeeded in separating the solubilized enzyme preparation by DEAE-cellulose chromatography into three components which were identified as cytochrome P-450, NADPH-cytochrome c reductase, and a heat-stable factor, respectively. All three components were required for the reconstitution of NADPH-dependent oxygenase activity. The solubilized and reconstituted preparations were also active in oxidizing various drugs (ethylmorphine, hexobarbital, aminopyrine, etc.).[101] Although the separated components were not yet pure at this time, Lu and Coon's successful solubilization and reconstitution of the microsomal oxygenase system provided a new approach for elucidating the nature and function of this important enzyme system.

The heat-stable factor was soon identified as phosphatidyl choline.[102] The lipid was apparently not involved in the oxygenation reaction itself, but was required for the association of the dispersed cytochrome P-450 and NADPH-cytochrome c reductase molecules to form an active enzyme aggregate. Later studies[103] confirmed the requirement for lipid in the reconstitution of the oxygenase activity from solubilized lipid-free components. In some cases, lipid could be replaced by a suitable neutral detergent.[104]

The purification and characterization of solubilized cytochrome P-450 and NADPH-cytochrome c reductase were actively pursued in subsequent years, and homogeneous preparations of both components from which the oxygenase activity can be reconstituted are now obtainable. Since the purification of cytochrome P-450 is described in detail in chapter 3, developments in the purification of NADPH-cytochrome c reductase will be discussed only briefly here.

As mentioned above, the successful solubilization and purification of microsomal NADPH-cytochrome c reductase had already been reported by the early 1960's,[12-14] and the properties of the purified enzyme had also been examined in detail. However, all of the previous purification procedures involved solubilization of the enzyme by the digestion of microsomes with hydrolytic enzymes, and the reductase thus solubilized from the microsomes was incapable of reducing cytochrome P-450 by NADPH.[100,102,105] We are now sure that a proteolytic process must always have been involved in the solubilization of the reductase by any of the earlier purification procedures, although some of them claimed the use of a purified lipase instead of a proteolytic enzyme.[106] Ichihara et al.[107] demonstrated that the reductase solubilized from microsomes with a detergent was active in reconstituting the oxygenase system with solubilized cytochrome P-450, but treatment of the detergent-solubilized reductase with trypsin resulted in a complete loss of the capability of the reductase to transfer reducing equivalents from NADPH to cytochrome P-450 whereas its catalytic activity for reducing cytochrome c was not impaired by the protease treatment.

Obviously, the protease treatment of microsome-bound or detergent-solubilized NADPH-cytochrome c reductase resulted in proteolytic cleavage of the native reductase molecule into a major enzymatically active fragment and a minor portion which was not required in the transfer of electrons from NADPH to cytochrome c but was essential for the interaction of the enzyme with the hydrophobic cytochrome P-450 molecule. The molecular weight of the detergent-solubilized reductase, 79,000–80,000,[45,108,109] did not differ greatly from corresponding values reported for protease-solubilized reductase preparations, 71,000[45] or 68,800.[110] Hence, the molecular weight of the particular minor portion of the reductase molecule which was lost from the native molecule by proteolysis can be assessed at around ten thousand daltons. As the protease-solubilized reductase showed no affinity for microsomal membrane, the lost minor portion is possibly the hydrophobic portion by which the enzyme molecule attaches to the membrane[111] (cf. section 4.1.1).

Another important development in the study of microsomal NADPH-cytochrome c reductase was the finding by Iyanagi and Mason[112] of the presence of both FAD and FMN in the reductase molecule. Since this reductase had long been considered an FAD enzyme,[12,106] the presence of two different kinds of flavin was unexpected. However, this finding, which was first reported in 1973, has been fully confirmed by later studies;[113] homogeneous preparations of detergent-solubilized NADPH-cytochrome c reductase contained one mole each of FMN and FAD per protein molecule.[108,109] Kinetic studies on the oxidation-reduction of the flavins in the reductase have been carried out,[114] and the significance of the presence of the two non-identical flavin molecules in this reductase in transferring electrons from the reductase to cytochrome P-450 was suggested.[114]

New purification procedures for NADPH-cytochrome c reductase in which the use of hydrolytic enzymes is avoided and the reductase is solubilized from microsomes by detergents, have been developed in the past few years. A homogeneous preparation of detergent-solubilized reductase[108] which did not contain any cytochrome b_5 or NADH-cytochrome b_5 reductase, could effectively reconstitute NADPH-dependent benzphetamine oxidation activity when combined with a solubilized and purified cytochrome P-450 and phosphatidylcholine. This and other reconstitution studies[103] appear to exclude the participation of cytochrome b_5 in the NADPH-supported oxidation of drugs by the cytochrome P-450 system. However, a recent reconstitution experiment by Lu et al.[115] has shown a marked stimulation of NADPH-dependent hydroxylation of chlorobenzene when purified cytochrome b_5[44] was added to a system consisting of cytochrome P-450, NADPH-cytochrome c reductase and lipid. Since the same oxidation activity of another reconstituted system containing cytochrome P-448 (cf. section 4.1.4) instead of cytochrome P-450 was not stimulated[109] by the addition of cytochrome b_5, it still remains difficult to draw a general conclusion from the results of these experiments to decide whether cytochrome b_5 is or is not involved in NADPH-dependent drug oxidation reactions catalyzed by intact microsomes.

On the other hand, the participation of cytochrome b_5 in the NADH-linked oxygenase activities of cytochrome P-450 has clearly been shown by similar reconstitution studies. In a reconstituted system consisting of cytochrome P-448 (or P-450), NADH-cytochrome b_5 reductase, and lipid, the addition of cytochrome b_5 was the obligatory requirement for the NADH-supported hydroxylation of 3.4-benzopyrene.[94] NADPH-cytochrome c reductase was completely ineffective in supporting the NADH-dependent hydroxylation of this substrate. The role of cytochrome b_5 in mediating the electron transfer from NADH-cytochrome b_5 reductase to cytochrome P-450 was thus established.

Reconstitution experiments with solubilized and purified microsomal components have yielded and are still yielding valuable information on the constitution and mechanism of reaction of the microsomal cytochrome P-450 system. However, as already stated above, it is necessary to be aware of the limitations in the applicability of observations with reconstituted systems to native membrane-bound enzyme systems. The microsomal cytochrome P-450 system appears to be a dynamic enzyme system in which the distribution or the assembly of its membrane-bound components may change significantly according to physiological or non-physiological variations in the condition of the animals. Such changes in the constitution of the enzyme system are likely to be accompanied by large alterations in the flow of electrons from NADPH or NADH to the terminal component, cytochrome P-450. The physiological significance of results obtained by reconstitution experiments must always be assessed by comparison with observations on intact microsomal systems.

4.1.4. Substrate Specificity of the Hepatic Microsomal Cytochrome P-450 System

Liver microsomes catalyze the oxidation of various biological and non-biological lipophilic organic compounds in the presence of NAD(P)H and molecular oxygen,[116,117] and most of these oxidation reactions are catalyzed by the cytochrome P-450 system. The inhibition of the reaction by carbon monoxide and the reversal of the carbon monoxide inhibition by irradiation with 450 nm light are commonly utilized in assessing the participation of cytochrome P-450 in a given microsomal oxygenation reaction, although the extent of the carbon monoxide inhibition may vary considerably from one substrate to another[118] and some cytochrome P-450-catalyzed reactions are not strongly inhibited by carbon monoxide under usual experimental conditions.[119] Table 4.3 lists the types of reactions which have so far been shown to be catalyzed by the cytochrome P-450 system of liver microsomes.

Several other oxygenase-type enzymes or enzyme systems are also present in liver microsomes (Table 4.3), and each of these catalyzes more or less specific oxidative reactions of high physiological importance (desaturation of fatty acids, oxidative degradation of heme, etc.). However, as will be seen from Table 4.3 there is no doubt about the tremendous diversity of reactions catalyzed by the cytochrome P-450 system of liver microsomes. Due to this apparent low substrate specificity of cytochrome P-450, liver microsomes are capable of metabolizing various synthetic organic compounds taken into the animal body as drugs, food additives, etc.

Before the finding of cytochrome P-450, it was possible to assume that many separate enzymes, each of which had a definite narrow sub-

TABLE 4.3. NAD(P)H-dependent monooxygenase-type activities of liver microsomes

(A) Activities in which cytochrome P-450 participates

Activity	Substrates
Hydroxylation of aliphatic hydrocarbon residues	lauric acid, n-hexane, etc.
Hydroxylation of alicyclic rings	cyclohexane, etc.
Hydroxylation of aromatic rings	aniline, benzopyrene, etc.
Hydroxylation of steroids	cholesterol, testosterone, etc.
N-Hydroxylation of imines to oximes	2,4,6-trimethylacetophenoneimine, etc.
Oxidative dealkylation of N-alkylamines	aminopyrine, ethylmorphine, etc.
Oxidative dealkylation of O-alkyl compounds	codeine, 7-ethoxycoumarine, etc.
Oxidative dealkylation of S-alkyl compounds	6-methylmercaptopurine, etc.
Oxidative deamination of primary amines	amphetamine, etc.
Oxidation of thioethers to sulfoxides and sulfones	chlorpromazine, phenothiazine, etc.
Oxidative dehalogenation of halogenated hydrocarbons	p-fluoroaniline, etc.
Epoxidation of hydrocarbons and halogenated hydrocarbons	aldrin, etc.

(B) Activities independent of cytochrome P-450

Activity	Ref. for non-involvement of cytochrome P-450
N-Oxidation of secondary and tertiary amines	120
2,3-Epoxidation of squalene	28
Oxidative cleavage of protohemin to biliverdin	29
Desaturation of stearyl CoA to oleyl CoA	33

strate specificity, were participating in the oxidation of various kinds of drugs by liver microsomes. However, the identification of cytochrome P-450 as the common terminal oxidase in the microsomal metabolism of various drugs as well as steroids[18,19] appears to have excluded such a possibility since the same photochemical action spectrum with a maximum at 450 nm is obtained for the light reversal of the carbon monoxide inhibition of the metabolism of all these substrates. Nevertheless, experimental evidence suggesting a multiple nature for the microsomal cytochrome P-450 system has also been presented. To cite a few instances, Welch et al.[121] studied the hydroxylation of testosterone in the 6β, 7α, and 16α positions by liver microsomes, and found that one hydroxylation reaction could be selectively stimulated or inhibited without influencing the others. Many pharmacologists[117] who studied the induced increase in the drug-oxidizing activity of liver microsomes in drug-treated animals noticed that the activity of liver microsomes from 3-methylcholanthrene-treated animals was significantly different from the corresponding activities of normal and phenobarbital-treated animals as regards both substrate specificity and sensitivity towards various inhibitors. These observations appeared inconsistent with a unitary nature for microsomal cytochrome P-450 as the common terminal oxygen-activating component of the microsomal drug-oxidizing enzyme system for all the various substrates, drugs and steroids.

The first strong evidence for the presence of multiple molecular species of cytochrome P-450 in liver microsomes was provided in 1966 by the finding of Sladek and Mannering[122] that the cytochrome P-450 in liver microsomes of methylcholanthrene-treated rats could be distinguished spectrally from normal cytochrome P-450 by an alteration in the pH-dependent shift[123] of the ethyl isocyanide-difference spectrum of its reduced form. They concluded that treatment with this particular inducer caused the formation of a new species of cytochrome P-450 which they termed "cytochrome P_1-450". In the next year, Alvares et al.[124] reported that dithionite-reduced microsomes from methylcholanthrene-treated rats showed a carbon monoxide-difference spectrum with a maximum at 448 nm which was different from the corresponding spectrum of normal cytochrome P-450 by about 2 nm. After their finding, the term "cytochrome P-448" was frequently used to distinguish this methylcholanthrene-induced cytochrome P-450. Hildebrandt et al.[125] also investigated the spectral difference between phenobarbital-induced and methylcholanthrene-induced cytochrome P-450's, and found that these two forms differed from each other not only in the carbon monoxide-difference spectra of their reduced forms but also in the absolute absorption spectra of their oxidized and reduced forms.

Thus the presence of two spectrally distinguishable forms of cytochrome P-450, normal (phenobarbital-induced) and methylcholanthrene-induced, was established in liver microsomes. It was also possible to distinguish the methylcholanthrene-induced cytochrome from normal cytochrome P-450 by electron paramagnetic resonance spectroscopy: the former contained high spin iron whereas low spin iron was predominant in the latter.[126,127] However, it was not yet clear whether the two forms represented two distinct molecular species of cytochrome P-450 or two interconvertible physical states of a single molecular species. Some investigators[124,128] believed that cytochrome P-448 was a new molecular species synthesized under the influence of the inducer because the inhibitors of protein or RNA synthesis were able to prevent the appearance of cytochrome P-448 in rats treated with methylcholanthrene. On the other hand, since tight binding of methylcholanthrene or its metabolite to liver proteins was detected,[129] some other investigators[125] suspected that the binding of the inducer to pre-existing cytochrome P-450 may have altered its spectral properties. Since this problem was very important, not only for understanding the mechanism of induction of microsomal cytochrome P-450 systems by various drugs but also in elucidating the substrate specificity and physiological function of the cytochrome, it attracted the attention of many workers, and numerous papers for[126,130,132] and against[133,134] the existence of multiple molecular species of cytochrome P-450 in liver microsomes were published in a few years around 1970.

Separation of cytochrome P-448 from normal-type cytochrome P-450 and purification of each hemoprotein species were obviously the best way to settle the dispute. However, because of the difficulty in purifying these hemoproteins from microsomes (*cf.* section 3.1), studies along this line have not been successful until very recently. Supporting evidence for the presence of multiple molecular species of cytochrome P-450 in liver microsomes increased steadily in the first half of the 1970's, and, unexpectedly, newer evidence disclosed the existence of more than two forms of cytochrome P-450 in liver microsomes.

Titration of liver microsomes with potassium cyanide revealed[135] the presence of three distinct types of cytochrome P-450 having widely different affinities for the cyanide. Detailed spectral studies on the binding of various ligands with the reduced form of cytochrome P-450 suggested[136] the existence of four, two major and two minor, components of cytochrome P-450 in liver microsomes from normal rats. The acrylamide gel electrophoresis of detergent-solubilized liver microsomes in the presence of sodium dodecyl-sulfate showed three[137] or four[43] heme-containing protein bands, all of which were assigned to different forms of cytochrome P-450. The estimated molecular weights of these multiple cytochrome P-450's were all around 50,000.

Studies on the hydroxylation of steroids by liver microsomes also indicated the presence of novel cytochrome P-450 species. 7α-Hydroxylation of cholesterol by rat liver microsomes was shown to be catalyzed by a unique species of cytochrome P-450[138] which had a very short half life of 2 h.[139] The liver microsomes of female rats catalyzed carbon monoxide-sensitive 15β-hydroxylation of steroid sulfates,[140] whereas this activity was absent from liver microsomes of male rats. All of these experimental results suggested the presence in liver microsomes of at least a few species of cytochrome P-450 each of which catalyzed the hydroxylation of particular steroids at specific positions.

Although we still do not have detailed information on the molecular properties of these multiple forms of microsomal cytochrome P-450, it is almost certain that each separate form has a narrower substrate specificity than the broad specificity observed in liver microsomds (Table 4.3). Some forms of cytochrome P-450 may have overlapping substrate specificity with others, while the reactions catalyzed by the cytochrome P-450's which hydroxylate steroids at particular positions may be quite specific. The enzymological approach for separating microsomal cytochrome P-450's into individual species is expected to provide a decisive answer to this problem, and the recent report by Ryan et al.[141] of the successful purification of both cytochromes P-448 and P-450 from rat liver microsomes has actually demonstrated that this approach is now feasible. They[141] examined the substrate specificity of the two isolated forms of cytochrome P-450 using a reconstituted system with lipid and NADPH-cytochrome c reductase, and found that they exhibited different specificities for benzphetamine and benzopyrene. Cytochrome P-450 was highly active in the N-demethylation reaction of benzphetamine but almost inactive in hydroxylating benzopyrene. Cytochrome P-448 showed the reverse specificity towards the two substrates. Judging from another reconstitution experiment[142] in which cross-recombinations of partially purified NADPH-cytochrome c reductases and cytochrome P-450's obtained from phenobarbital-treated and methylcholanthrene-treated rats were examined, the reductase has nothing to do with the substrate specificity of the cytochrome P-450 system. The lipid fraction, which is also required for reconstitution of the oxygenase activity of cytochrome P-450, also does not contribute to the substrate specificity of the reconstituted system.[143]

Characterization of the multiple forms of cytochrome P-450 and elucidation of the substrate specificity of each individual form will undoubtedly contribute to our understanding of the physiological function of the microsomal cytochrome P-450 system. Since drugs, which are rapidly metabolized by this enzyme system, are taken into the animal body only on comparatively rare occasions, the normal function of this "drug-oxidizing" enzyme system has been the subject of controversy from the begining of studies on cytochrome P-450. The overlapping substrate specificities of some forms of the cytochrome, which cover endogenous metabolites (steroids, fatty acids, etc.) as well as various drugs, should provide a working answer to this question. Studies in this field are now progressing rapidly, so that a better understanding of the physiological significance of the "drug-oxidizing" enzyme system of liver microsomes should soon be forthcoming.

Finally, mention should be made of one more possible function of cytochrome P-450 in vivo, to which little attention has so far been paid. Since this hemoprotein has a very low redox potential, -0.34 to -0.40 V,[144] it should be capable of donating electrons to various electron acceptors if they can interact with the active portion of the cytochrome molecule. Such a reductive function for cytochrome P-450 would be manifested more distinctly under anaerobic conditions since the reduced form of this hemoprotein has a high affinity for molecular oxygen.[17,60] The participation of cytochrome P-450 in the NADPH-dependent reduction of azo-compounds[145] or nitro-compounds[146] by liver microsomes has already been suggested. A recent report by Sugiura et al.[147] on the NAD(P)H-supported reduction of tertiary amine N-oxide confirms more clearly the participation of cytochrome P-450 in this reductive reaction. The postulated electron flow from NADH to cytochrome P-450 via cytochrome b_5 (section 4.1.2) suggests that the reverse reaction, the donation of reducing equivalents from cytochrome P-450 to cytochrome b_5, should also be feasible. Since the partial pressure of oxygen in the liver cell is much lower than that in air-saturated aqueous solutions, the reductive function of cytochrome P-450 which is not so marked under usual *in vitro* experimental conditions, may have far more significance *in vivo*. This non-oxidative function of cytochrome P-450 must be re-examined in the future.

REFERENCES

Section 4.1
1. G. H. Hogeboom, *J. Biol. Chem.*, **177**, 847 (1949).
2. G. H. Hogeboom and W. C. Schneider, *ibid.*, **186**, 417 (1950).
3. G. E. Palade and P. Siekevitz, *J. Biophys. Biochem. Cytol.*, **2**, 171 (1956).
4. H. Beinert, *J. Biol. Chem.*, **190**, 287 (1951).
5. W. C. Schneider, *ibid.*, **165**, 585 (1946).
6. G. H. Hogeboom, A. Claude and R. D. Hotchkiss, *ibid.*, **165**, 615 (1946).
7. C. de Duve, B. C. Pressman, R. Gianetto, R. Wattiaux and F. Appelmans, *Biochem. J.*, **60**, 604 (1955).
8. C. F. Strittmatter and E. G. Ball, *Proc. Natl. Acad. Sci. U.S.A.*, **38**, 19 (1952).
9. H. Yoshikawa, *J. Biochem. (Tokyo)*, **38**, 1 (1951).
10. P. Strittmatter and S. F. Velick, *J. Biol. Chem.*, **221**, 253 (1956).
11. P. Strittmatter and S. F. Velick, *ibid.*, **221**, 277 (1956).
12. C. H. Williams, Jr. and H. Kamin, *ibid.*, **237**, 587 (1962).
13. A. H. Phillips and R. G. Langdon, *ibid.*, **237**, 2652 (1962).
14. B. L. Horecker, *ibid.*, **183**, 593 (1950).
15. H. Nishibayashi, T. Omura and R. Sato, *Biochim. Biophys. Acta*, **67**, 520 (1963).
16. T. Omura and R. Sato, *J. Biol. Chem.*, **237**, 1375 (1962).
17. T. Omura and R. Sato, *ibid.*, **239**, 2370 (1964).
18. R. W. Estabrook, D. Y. Cooper and O. Rosenthal, *Biochem. Z.*, **338**, 741 (1963).
19. D. Y. Cooper, S. Levin, S. Narasimhulu, O. Rosenthal and R. W. Estabrook, *Science*, **147**, 400 (1965).
20. M. Klingenberg, *Arch. Biochem. Biophys.*, **75**, 376 (1958).
21. D. Garfinkel, *ibid.*, **77**, 493 (1958).
22. Y. Hashimoto, T. Yamano and H. S. Mason, *J. Biol. Chem.*, **237**, 3843 (1962).
23. Y. Ichikawa and T. Yamano, *Arch. Biochem. Biophys.*, **121**, 742 (1967).
24. Y. Ichikawa, B. Hagihara and T. Yamano, *ibid.*, **120**, 204 (1967).
25. T. Omura and S. Takesue, *Seikagaku* (Japanese), **40**, 608 (1968).

26. B. S. Cohen and R. W. Estabrook, *Arch. Biochem. Biophys.*, **143**, 37 (1971).
27. L. Ernster and S. Orrenius, *Fed. Proc.*, **24**, 1190 (1965).
28. T. Ono and K. Bloch, *J. Biol. Chem.*, **250**, 1571 (1975).
29. T. Yoshida, S. Takahashi and G. Kikuchi, *J. Biochem. (Tokyo)*, **75**, 1187 (1974).
30. D. W. Lubbers, M. Kessler, R. Scholz and T. Bucher, *Biochem. Z.*, **341**, 346 (1965).
31. P. W. Holloway and S. J. Wakil, *J. Biol. Chem.*, **245**, 1862 (1970).
32. N. Oshino, Y. Imai and R. Sato, *J. Biochem. (Tokyo)*, **69**, 155 (1971).
33. N. Oshino, Y. Imai and R. Sato, *Biochim. Biophys. Acta*, **128**, 13 (1966).
34. P. W. Holloway and J. T. Katz, *Biochemistry*, **11**, 3689 (1972).
35. T. Shimakata, K. Mihara and R. Sato, *J. Biochem. (Tokyo)*, **72**, 1163 (1972).
36. N. Oshino and T. Omura, *Arch. Biochem. Biophys.*, **157**, 395 (1973).
37. F. Paltauf, R. A. Prough, B. S. S. Masters and J. M. Johnson, *J. Biol. Chem.*, **249**, 2661 (1974).
38. C. F. Strittmatter and E. G. Ball, *J. Cell. Comp. Physiol.*, **43**, 57 (1954).
39. P. Siekevitz, *Ann. Rev. Physiol.*, **25**, 15 (1963).
40. G. Dallner, P. Siekevitz and G. E. Palade, *J. Cell. Biol.*, **30**, 97 (1966).
41. S. J. Singer and G. L. Nicolson, *Science*, **175**, 720 (1972).
42. A. Ito and R. Sato, *J. Biol. Chem.*, **243**, 4922 (1968).
43. A. P. Alvares and P. Siekevitz, *Biochem. Biophys. Res. Commun.*, **54**, 923 (1973).
44. L. Spatz and P. Strittmatter, *Proc. Natl. Acad. Sci. U.S.A.*, **68**, 1042 (1971).
45. A. F. Welton, T. C. Pederson, J. A. Buege and S. D. Aust, *Biochem. Biophys. Res. Commun.*, **54**, 161 (1973).
46. L. Spatz and P. Strittmatter, *J. Biol. Chem.*, **248.**, 793 (1973).
47. P. Strittmatter, M. J. Rogers and L. Spatz, *ibid.*, **247**, 7188 (1972).
48. K. Enomoto and R. Sato, *Biochem. Biophys. Res. Commun.*, **51**, 1 (1973).
49. T. Omura, P. Siekevitz and G. E. Palade, *J. Biol. Chem.*, **242**, 2389 (1967).
50. A. Ito and R. Sato, *J. Cell. Biol.*, **40**, 179 (1969).
51. Y. Kuriyama, T. Omura, P. Siekevitz and G. E. Palade, *J. Biol. Chem.*, **244**, 2017 (1969).
52. S. Takesue and T. Omura, *Biochem. Biophys. Res. Commun.*, **40**, 396 (1970).
53. M. J. Rogers and P. Strittmatter, *J. Biol. Chem.*, **249**, 895 (1974).
54. G. Dallner, A. Bergstrand and R. Nilsson, *J. Cell. Biol.*, **38**, 257 (1968).
55. P. R. Dallman, G. Dallner, A. Bergstrand and L. Ernster, *ibid.*, **41**, 357 (1969)
56. H. U. Schulze, J. M. Ponnighaus and H. Staudinger, *Z. Physiol. Chem.*, **353**, 1195 (1972).
57. M. R. Franklin and R. W. Estabrook, *Arch. Biochem. Biophys.*, **143**, 318 (1971).
58. A. Ito, *J. Biochem. (Tokyo)*, **75**, 787 (1974).
59. A. Stier and E. Sackmann, *Biochem. Biophys. Acta*, **311**, 400 (1973).
60. T. Omura, R. Sato, D. Y. Cooper, O. Rosenthal and R. W. Estabrook, *Fed. Proc.*, **24**, 1181 (1965).
61. P. L. Gigon, T. E. Gram and J. R. Gillette, *Mol. Pharmacol.*, **5**, 109 (1969).
62. Y. Miyake, K. Mori and T. Yamano, *Biochem. Biophys. Res. Commun.*, **44**, 564 (1971).
63. B. N. La Du, L. Gandette, N. Trousof and B. B. Brodie, *J. Biol. Chem.*, **214**, 741 (1955).
64. K. Krisch and H. Staudinger, *Biochem. Z.*, **334**, 312 (1961).
65. T. Omura and R. Sato, *J. Biol. Chem.*, **239**, 2379 (1964).
66. R. Sato, T. Omura and H. Nishibayashi, *Oxidases and Related Redox Systems* (ed. T. E. King, H. S. Mason, and M. Morrison), vol. 1, p. 861, John Wiley (1965).
67. S. Orrenius, A. Berg and L. Ernster, *Eur. J. Biochem.*, **11**, 193 (1969).
68. T. Omura, *Microsomes and Drug Oxidations* (ed. J. R. Gillette, A. H. Conney, G. J. Cosmides, R. W. Estabrook, J. R. Fouts and G. J. Mannering), p. 160, Academic Press (1969).
69. F. Wada, H. Shibata, M. Goto and Y. Sakamoto, *Biochim. Biophys. Acta*, **162**, 518 (1968).

70. R. I. Glazer, J. B. Schenkman and A. C. Sartorelli, *Mol. Pharmacol.*, **7**, 683 (1971).
71. B. S. S. Masters, J. Baron, W. E. Taylor, E. L. Isaacson and J. L. Spalluto, *J. Biol. Chem.*, **246** 4143 (1971).
72. M. Raftell and S. Orrenius, *Biochim. Biophys. Acta*, **233**, 358 (1971).
73. A. Y. H. Lu and M. J. Coon, *J. Biol. Chem.*, **243**, 1331 (1968).
74. T. Omura, E. Sanders, R. W. Estabrook, D. Y. Cooper and O. Rosenthal, *Arch. Biochem. Biophys.*, **117**, 660 (1966).
75. M. Katagiri, B. N. Ganguli and I. C. Gunsalus, *J. Biol. Chem.*, **243**, 3543 (1968).
76. Y. Miyake, J. L. Gaylor and H. S. Mason, *ibid.*, **243**, 5788 (1968).
77. I. Hoffstrom, A. Ellin, S. Orrenius, D. Backstrom and A. Ehrenberg, *Biochem. Biophys. Res. Commun.*, **48**, 977 (1972).
78. J. Peisach and W. E. Blumberg, *Proc. Natl. Acad. Sci. U.S.A.*, **67**, 172 (1970).
79. T. A. van der Hoeven and M. J. Coon. *J. Biol. Chem.*, **249**, 6302 (1974).
80. A. H. Conney, R. R. Brown, J. A. Miller and E. C. Miller, *Cancer Res.*, **17**, 628 (1957).
81. A. Nilsson and B. C. Johnson, *Arch. Biochem. Biophys.* **101**, 494 (1963).
82. Y. Ichikawa, T. Yamano and H. Fujisawa, *Biochim. Biophys. Acta*, **171**, 32 (1969).
83. Y. Ichikawa and T. Yamano, *J. Biochem. (Tokyo)*, **66**, 351 (1969).
84. Y. Ichikawa and J. S. Loehr, *Biochem. Biophys. Res. Commun.*, **46**, 1187 (1972).
85. Y. Ichikawa and T. Yamano, *Biochim. Biophys. Acta*, **200**, 220 (1970).
86. B. S. Cohen and R. W. Estabrook, *Arch. Biochem. Biophys.*, **143**, 46 (1971).
87. B. S. Cohen and R. W. Estabrook, *ibid.*, **143**, 54 (1971).
88. R. W. Estabrook, A. Hildebrandt, J. Baron, K. J. Netter and K. Leibman, *Biochem. Biophys. Res. Commun.*, **42**, 132 (1971).
89. A. Hildebrandt and R. W. Estabrook, *Arch. Biochem. Biophys.*, **143**, 66 (1971).
90. M. A. Correia and G. J. Mannering, *Mol. Pharmacol.*, **9**, 455 (1973).
91. M. A. Correia and G. J. Mannering, *ibid.*, **9**, 470 (1973).
92. I. Jansson and J. B. Schenkman, *ibid.*, **9**, 840 (1973).
93. H. Staudt, F. Lichtenberger and V. Ullrich, *Eur. J. Biochem.*, **46**, 99 (1974).
94. S. B. West, W. Levin, D. Ryan, M. Vore and A. Y. H. Lu, *Biochem. Biophys. Res. Commun.* **58**, 516 (1974).
95. A. Y. H. Lu, S. B. West, M. Vore, D Ryan and W. Levin, *J. Biol. Chem.*, **249**, 6701 (1974).
96. H. A. Sasame, J. R. Mitchell, S. S. Thorgeirsson and J. R. Gillette, *Drug Metab. Disposition*, **1**, 121 (1973).
97. G. J. Mannering, S. Kuwahara and T. Omura, *Biochem. Biophys. Res. Commun.* **57**, 476 (1974).
98. H. A. Sasame, S. S. Thorgeirsson, J. R. Mitchell and J. R. Gillette, *Life Sci.*, **14**, 35 (1974).
99. Y. Ichikawa and T. Yamano, *Biochim. Biophys. Acta*, **131**, 490 (1967).
100. A. Y. H. Lu, K. W. Junk and M. J. Coon, *J. Biol. Chem.*, **244**, 3714 (1969).
101. A. Y. H. Lu, H. W. Strobel and M. J. Coon, *Biochem. Biophys. Res. Commun.*, **36**, 545 (1969).
102. H. W. Strobel, A. Y. H. Lu, J. Heidema and M. J. Coon, *J. Biol. Chem.*, **245**, 4851 (1970).
103. W. Levin, D. Ryan, S. West and A. Y. H. Lu, *ibid.*, **249**, 1747 (1974).
104. K. Ichihara, E. Kusunose and M. Kusunose, *Biochim. Biophys. Acta*, **239**, 178 (1971).
105. A. Y. H. Lu, K. W. Junk and M. J. Coon, *Mol. Pharmacol.*, **6**, 213 (1970).
106. B. S. S. Masters, C. H. Williams, Jr. and H. Kamin, *Methods in Enzymol.*, **10**, 565 (1970).
107. K. Ichihara, E. Kusunose and M. Kusunose, *Eur. J. Biochem.*, **38**, 463 (1973).
108. J. L. Vermilion and M. J. Coon, *Biochem. Biophys. Res. Commun.*, **60**, 1315 (1974).
109. J. D. Dignam and H. W. Strobel, *ibid.*, **63**, 845 (1975).
110. B. S. S. Masters and D. M. Ziegler, *Arch. Biochem. Biophys.*, **145**, 358 (1971).
111. H. Satake and R. Sato, *Seikagaku* (Japanese), **45**, 582 (1973).

HEPATIC MICROSOMAL SYSTEMS 163

112. T. Iyanagi and H. S. Mason, *Biochemistry*, **12**, 2297 (1973).
113. T. A. van der Hoeben and M. J. Coon, *J. Biol. Chem.*, **249**, 6302 (1974).
114. T. Iyanagi, N. Makino and H. S. Mason, *Biochemistry*, **13**, 1701 (1974).
115. A. Y. H. Lu, W. Levin, H. Salander and D. M. Jerina, *Biochem. Biophys. Res. Commun.*, **61**, 1348 (1974).
116. B. B. Brodie, J. R. Gillette and B. N. La Du, *Ann. Rev. Biochem.*, **27**, 427 (1958).
117. A. H. Conney, *Pharmacol. Rev.* **19**, 317 (1967).
118. A. H. Conney, W. Levin, M. Ikeda, R. Kuntzman, D. Y. Cooper and O. Rosenthal, *J. Biol. Chem.*, **243**, 3912 (1968).
119. E. A. Thompson, Jr. and P. K. Siiteri, *ibid.*, **249**, 5373 (1974).
120. D. M. Ziegler and C. H. Mitchell, *Arch. Biochem. Biophys.*, **150**, 116 (1972).
121. R. M. Welch, W. Levin and A. H. Conney, *J. Pharmacol. Exptl. Ther.*, **155**, 225 (1964).
122. N. E. Sladek and G. J. Mannering, *Biochem. Biophys. Res. Commun.*, **24**, 668 (1966).
123. Y. Imai and R. Sato, *ibid.*, **23**, 5 (1966).
124. A. P. Alvares, G. Schilling, W. Levin and R. Kuntzman, *ibid.*, **29**, 521 (1967).
125. A. Hildebrandt, H. Remmer and R. W. Estabrook, *ibid.*, **30**, 607 (1968).
126. C. R. E. Jefcoate and J. L. Gaylor, *Biochemistry*, **8**, 3464 (1969).
127. C. R. E. Jefcoate, R. L. Calabrese, and J. L. Gaylor, *Mol. Pharmacol.*, **6**, 391 (1970).
128. A. P. Alvares, G. Schilling, W. Levin and R. Kuntzman, *J. Pharmacol. Exptl. Ther.*, **163**, 417 (1968).
129. E. Bresnick, R. A. Liebelt, J. G. Stevenson and J. C. Madix, *Cancer Res.*, **27**, 462 (1967).
130. K. Einarsson and G. Johansson, *Eur. J. Biochem.*, **6**, 293 (1968).
131. W. Levin and R. Kuntzman, *J. Biol. Chem.*, **244**, 3671 (1969).
132. F. J. Wiebel, J. C. Leutz, L. Diamond and H. V. Gelboin, *Arch. Biochem. Biophys.*, **144**, 78 (1971).
133. J. B. Schenkman, H. Greim, M. Zange and H. Remmer, *Biochim. Biophys. Acta*, **171**, 23 (1969).
134. Y. Imai and P. Siekevitz, *Arch. Biochem. Biophys.*, **144**, 143 (1971).
135. K. Comai and J. L. Gaylor, *J. Biol. Chem.*, **248**, 4947 (1973).
136. J. Werringloer and R. W. Estabrook, *Arch. Biochem. Biophys.*, **167**, 270 (1975).
137. A. F. Welton and S. D. Aust, *Biochem. Biophys. Res. Commun.*, **56**, 898 (1974).
138. G. S. Boyd, A. M. Grimwade and M. E. Lawson, *Eur. J. Biochem.*, **37**, 334 (1973).
139. J. Gielen, J. Van Cantfort, B. Robaye and J. Renson, *ibid.*, **55**, 41 (1975).
140. J. A. Gustafsson and M. Ingelman-Sundberg, *J. Biol. Chem.*, **249**, 1940 (1974).
141. D. Ryan, A. Y. H. Lu, J. Kawalek, S. B. West, and W. Levin, *Biochem. Biophys. Res. Commun.*, **64**, 1134 (1975).
142. T. Fujita and G. J. Mannering, *J. Biol. Chem.*, **248**, 8150 (1973).
143. D. W. Nebert, J. K. Heidema, H. W. Strobel and M. J. Coon, *ibid.*, **248**, 7631 (1973).
144. M. R. Waterman and H. S. Mason, *Arch. Biochem. Biophys.*, **150**, 57 (1972).
145. P. H. Hernandez, J. R. Gillette and P. Mazel, *Biochem. Pharmacol.*, **16**, 1859 (1967).
146. J. R. Gillette, J. J. Kamm and H. A. Sasame, *Mol. Pharmacol.*, **4**, 541 (1968).
147. M. Sugiura, K. Iwasaki, and R. Kato *ibid.*, **12**, 322 (1976).

4.2 Adrenocortical Mitochondrial Systems

The cholesterol side-chain cleavage enzyme, the steroid 11β-hydroxylase, and the steroid 18-hydroxylase are known to occur in adrenocortical mitochondria. The hydroxylation of deoxycorticosterone at the carbon 11β position requires the presence of reduced pyridine nucleotide and molecular oxygen.[1,2] This reaction has been studied by the incorporation of ^{18}O from molecular oxygen into the substrate molecule, deoxycorticosterone,[3,4] and on the basis of these observations it has been suggested that 11β-hydroxylase belongs to the class of enzymes known as mixed-function oxidases. Harding et al.[5] first reported the presence of cytochrome P-450 in adrenocortical mitochondria and this finding was later supported by other workers.[6-8] Cooper et al.[9] demonstrated that carbon monoxide inhibition of the 11β-hydroxylase enzyme system was light reversible and that the optimum wavelength for reversal of the CO-inhibition was 450 nm. These findings led to the conclusion that cytochrome P-450 serves as a mixed-function oxidase for the 11β-hydroxylation of deoxycorticosterone. The hydrolase system has been shown to consist of three protein components, which have been identified as a flavoprotein (NADPH-adrenodoxin reductase), an iron-sulfur protein (adrenodoxin) and a hemoprotein (cytochrome P-450).[10-12] The cholesterol side-chain cleavage system of adrenocortical mitochondria has been shown by Halkerston et al.[13] to belong to the mixed-function oxidase requiring NADPH and molecular oxygen. Simpson and Boyd[14,15] have further demonstrated the requirement of similar electron transfer components to the 11β-hydroxylase enzyme system. The electron transfer system of adrenal mitochondria is different from the liver microsomal type system, which has been resolved into a flavoprotein (NADPH-cytochrome P-450 reductase) and a hemoprotein (cytochrome P-450), but does not contain an iron-sulfur protein.[16]

In order to provide more detailed information on the reaction mechanisms of the steroid hydroxylase systems in adrenocortical mitochondria, it is valuable that the three functionally active protein components be available in a pure state and that the characterizations of each component be investigated at the enzymological level. In this section, both the purification and characterizations of the three components and the reaction mechanisms of the steroid hydroxylase enzyme systems in adrenocortical mitochondria, will be described.

4.2.1 Sequential Fragmentation of Steroid Hydroxylating Components from Adrenal Cortex

The procedure outlined here is favored especially for the simultaneous isolation of all components of the adrenal steroid hydroxylase system; the reductase, the ferredoxin and the cytochrome P-450.[17] The frozen adrenal cortexes are thawed and homogenized with 250 mM sucrose solution. This homogenate is centrifuged at 700g for 10 min, and the cloudy supernatant is decanted off and saved. The precipitate is then homogenized again and centrifuged as above. The supernatants from the first and second centrifugations are combined and centrifuged at 9000g for 30 min. The supernatant (S_1) is saved. The sedimented pellets are then suspended in the sucrose solution and the differential centrifugation (i.e. 700g for 10 min and 9000g for 30 min) is repeated. The supernatant (S_2) is saved. Most of the adrenodoxin is extracted into the S_1 and S_2 fractions. After washing with 100 mM potassium phosphate buffer (pH 7.0), the mitochondrial pellets are suspended in the buffer and sonicated. After centrifugation of the mixture, the resultant supernatant (S_3) is used for the isolation of NADPH-adrenodoxin reductase, while the mitochondrial pellets (P_1) serve for the extraction of cytochrome P-450.

4.2.2 Adrenodoxin

A. Purification

Isolation and purification of porcine and bovine adrenodoxins have been carried out by Kimura and Suzuki[12,18] and Omura et al.,[11] who obtained the adrenodoxin from either adrenal mitochondria or adrenal gland as starting material. As described in section 4.2.1, adrenodoxin can easily be extracted into the sucrose solution (S_1 and S_2) during the fragmentation of mitochondria from adrenal cortex.[17] The S_1 and S_2 fractions containing adrenodoxin are adsorbed on a large column (5×25 cm) of DEAE-cellulose. The eluated preparation is then twice subjected to chromatography on a DEAE-cellulose column (2.8×30 cm), followed by gel filtration on Sephadex G-100 (2.8×70 cm). The final preparation is further crystallized. On additon of solid ammonium sulfate, brownish-red crystals of adrenodoxin appear as needles (Fig. 4.2). The yield of twice-recrystallized adrenodoxin is about 60 mg from 1 kg of adrenal cortex. A summary of the purification of one such preparation of adrenodoxin from the S_1 and S_2 fractions is given in Table 4.4. The purity index ($A_{414}\text{nm}/A_{276}\text{nm}$) was found to

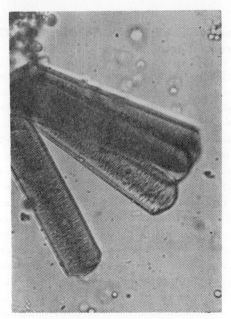

Fig. 4.2. Crystals of bovine adrenodoxin.

TABLE 4.4. Purification of adrenodoxin from bovine adrenal cortex (1 kg)

Step	Volume (ml)	Total protein (mg)	Purity index (A_{414} nm/A_{276} nm)
First DEAE-cellulose chromatography	440	2,600	†
Third DEAE-cellulose chromatography	260	230	0.46
Sephadex G-100 gel filtration	50	123	0.73
First crystals	2.4	80	0.82
Second crystals	1.7	60	0.86

† Omitted because the preparation at this stage contains a large amount of colored impurities.

be 0.86, which is higher than the value of 0.76 previously reported.[12] The final preparation appears to be homogeneous as judged from polyacrylamide gel electrophoresis and ultracentrifugration.

Apo-adrenodoxin has been obtained by treatment of the native protein with an iron-chelating agent such as α,α'-dipyridyl[18] or with trichloroacetic acid.[19] The procedure most commonly used, however, is treatment of the native protein with trichloroacetic acid (5% final concentration). When adrenodoxin is so treated with trichloroacetic acid, the color of the protein is bleached, and the apo-protein formed is devoid of enzymatic activity. The apo-protein contains no significant amounts of iron (less than 5 ng iron atoms/mg protein) or labile sulfide (less than 2 nmole/mg protein). The apo-protein thus prepared can recombine with Fe^{2+} and sulfide to yield adrenodoxin. However, the reconstitution appears to be incomplete,[18] and the yield of reconstituted protein is considerably low (20%). Suhara et al.[20] have investigated the detailed conditions necessary for maximum conversion of the apo-adrenodoxin to the reconstituted form. The recombination reaction of the apo-protein with Fe^{2+} and sulfide requires that the apo-protein be kept in the presence of a high concentration of urea. The yield can be as high as 75%. Excess reagent and contaminated inactive apo-protein are removed by dialysis and a DEAE-cellulose chromatography step. The reconstituted adrenodoxin is then finally crystallized. The crystals are quite similar to those of native adrenodoxin.

B. Properties

The absorption spectrum of adrenodoxin is shown in Fig. 4.3. The oxidized adrenodoxin exhibits absorption maxima at 414 and 455 nm with a broad maximum at 320 nm. The hyperfine structure is observed in the ultraviolet region, suggesting that a perturbation of electron densities is associated with an aromatic amino acid near the iron-sulfur region of the molecule. In contrast to the native protein, the apo-protein spectrum exhibits a maximum in the ultraviolet region, but no absorption in the visible region. The reconstituted holo-protein appears to have identical spectral properties to native adrenodoxin. Reduction of adrenodoxin, either enzymatically by NADPH via adrenodoxin reductase or chemically with sodium dithionite, gives rise to about a half decolorization of the visible absorption between 400 and 500 nm, concomitant with the appearance of a broad absorption at 540 nm. If the absorbance is entirely attributable to the oxidized chromophore,

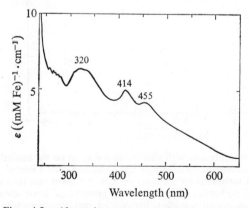

Fig. 4.3. Absorption spectrum of bovine adrenodoxin.

a half reduction of adrenodoxin must occur in such reactions.[18,21]

The iron contents of the crystalline native and reconstituted preparations are both estimated to be about 120 ng iron atoms/mg protein, as determined by a biuret reaction. This corresponds to about 150 ng iron atoms/mg dry weight of protein. When the labile sulfide of adrenodoxin is assayed by the original method of Fogo and Popowsky,[22] a value of 58 nmole per mg of protein is obtained. However, as shown in Fig. 4.4, an extension of the incubation time with alkaline zinc acetate solution to over 2 h results in the release of a maximal amount of sulfide: 120 nmole labile sulfide per mg of protein. Thus, the molar ratio of iron to sulfur approaches one.[20] The additional detectable labile sulfide is not due to H_2S derived directly from cysteinyl residues by a β-elimination reaction during the prolonged incubation with alkaline zinc acetate.[23]

The amino acid sequence of bovine adrenodoxin has been determined by Tanaka et al.[24,25] The molecule is composed of 114 amino acids in the form of a single polypeptide chain as shown in Fig. 4.5. The fine structure of adrenodoxin is not yet well understood and further structural information will depend on the results of X-ray crystallographic work. Other properties of adrenodoxin are summarized in Table 4.5.

The electron transfer function of adrenodoxin can be examined by measuring its stimulatory effect on the rate of cytochrome c reduction in the presence of NADPH-adrenodoxin reductase and NADPH. The specific activity has been found to be 116 mkat/kg protein. When adre-

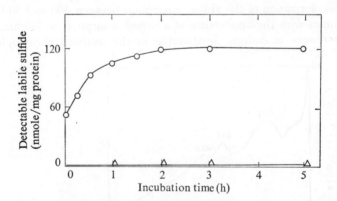

Fig. 4.4. Effect of incubation with alkaline zinc reagent on the analysis of the labile sulfide of adrenodoxin. The sample was incubated with a mixture of zinc acetate and sodium hydroxide for the specified time indicated and then coupled with N,N'-dimethyl-p-phenylenediamine in the presence of ferric chloride to give methylene blue, which was then determined spectrophotometrically. ○, Adrenodoxin; △, apo-adrenodoxin.

Ser 1	Ser 2	Ser 3	Gln 4	Asp 5	Lys 6	Ile 7	Thr 8	Val 9	His 10
Phe 11	Ile	Asn	Arg	Asp	Gly	Glu	Thr	Leu	Thr 20
Thr 21	Lys	Gly	Lys	Ile	Gly	Asp	Ser	Leu	Leu 30
Asp 31	Val	Val	Val	Gln	Asn	Asn	Leu	Asp	Ile 40
Asp 41	Gly	Phe	Gly	Ala	Cys	Glu	Gly	Thr	Leu 50
Ala 51	Cys	Ser	Thr	Cys	His	Leu	Ile	Phe	Glu 60
Gln 61	His	Ile	Phe	Glu	Lys	Leu	Glu	Ala	Ile 70
Thr 71	Asn	Glu	Glu	Asn	Asn	Met	Leu	Asp	Leu 80
Ala 81	Tyr	Gly	Leu	Thr	Asp	Arg	Ser	Arg	Leu 90
Gly 91	Cys	Gln	Ile	Cys	Leu	Thr	Lys	Ala	Met 100
Asp 101	Asn	Met	Thr	Val	Arg	Val	Pro	Asp	Ala 110
Val 111	Ser	Asp	Ala 114						

Fig. 4.5. Primary structure of bovine adrenodoxin.

TABLE 4.5. Physicochemical properties of bovine adrenodoxin

Properties	Value
Molecular weight by	
sedimentation equilibrium[12]	12,000
sedimentation-diffusion[12]	11,400
amino acid composition[24]	12,638
Fe, sulfide content[17]	13,300
$S_{20,w}$[20]	1.65 S (native protein)
	1.45 S (apoprotein)
	1.83 S (reconstituted protein)
$D_{20,w}$[12]	11×10^{-7} cm²/sec
Fe and S^{2-} contents per mole of protein[12]	2
Absorbance indices[17]	
nm ε ((M Fe^{-1}), cm^{-1})	K ((g protein/ml)$^{-1}$, cm^{-1})
276 5800	675
320 6400	—
414 5000	578
455 4200	—
EPR spectrum of reduced form[26]	$g = 1.94$
E_0[12]	164 mV
Mole of reducing equivalent per mole of protein[12]	1
ORD[12]	determined
CD[18]	determined
Magnetic susceptibility[18]	diamagnetic (oxidized form)

nodoxin is reduced with NADPH and adrenodoxin reductase, superoxide anion is generated, as judged by the accumulation of adrenochrome in the presence of epinephrine and O_2. The reaction is greatly inhibited by superoxide desmutase. Since each adrenochrome corresponds to

1.39 O_2^-, as reported,[27] the rate of generation of superoxide anion can be estimated to be 1.5 kat/mole adrenodoxin.[20]

4.2.3 NADPH-Adrenodoxin Reductase

A. Purification

NADPH-adrenodoxin reductase has been isolated and partially purified from adrenal mitochondria of beef,[11] pig[12] and rat.[10] However, the low yield of this enzyme has hampered its further purification. During the course of isolation of mitochondria from adrenal cortexes, it was found that NADPH-adrenodoxin reductase was partially extracted into the sucrose supernatant fraction (S_1 and S_2), about half of the total activity being extractable. Attempts to purify the reductase from such sucrose extracts have proved impractical, since the sucrose extract has a low starting specific activity and the activity is rapidly lost during purification. Reductase activity is also found in both the soluble and pellet fractions obtained after sonication of the mitochondrial fraction. When the pellet fractions are dispersed with 1.5% Triton X-100, about 15% of the total reductase activity is solubilized. The preparation can be purified partially by means of ammonium sulfate fractionation followed by DEAE-cellulose chromatography. However, in this case, either the specific activity or the yield is low. Considering the content, specific activity and location of the two other components of the system, adrenodoxin (the sucrose extract) and cytochrome P-450 (the pellets after sonication), the sonicated supernatant represents a better source material for the purification of NADPH-adrenodoxin reductase.[28,29] It contains about 35% of the total reductase activity.

The yellow-brown supernatant (S_3) obtained after sonication of mitochondrial pellets is fractionated at between 35 and 60% saturation of ammonium sulfate. After dialysis, the reductase is applied to a DEAE-cellulose column (2.8×30 cm) equilibrated in 10 mM potassium phosphate buffer, and the reductase is then eluted with 50 mM buffer. After ammonium sulfate fractionation (35 and 55% saturation), the dialyzed sample is again applied to a DEAE-cellulose column and then eluted with a linear gradient between 10 mM and 50 mM buffer. The chromatographic pattern of the reductase is usually separated into two fractions. The major active fractions are combined and fractionated with ammonium sulfate (40 and 55% saturation). The reductase is passed through a Sephadex G-100 column (2.8×60 cm) equilibrated in 10 mM buffer containing 100 mM KCl. After concentration with ammonium sulfate, the preparation can be stored at −80°C for a few weeks without substantial loss of activity. The yield is about 10 mg

TABLE 4.6. Summary of the purification of NADPH-adrenoxin reductase†

Fraction	Volume (ml)	Protein (mg)	Activity (μkat)	Specific activity (mkat/mg protein)	Yield (%)
Crude extract	580	5836	16.2	2.8	100
Ammonium sulfate	50	2000	8.3	4.2	52
1st DEAE-cellulose	10	170	4.8	28	30
2nd DEAE-cellulose	2.3	35	1.6	46	10
Sephadex G-100	1.5	19	1.2	63	7.2

† The activity of the reductase is routinely assayed by measuring the rate of adrenodoxin-dependent reduction of cytochrome c.

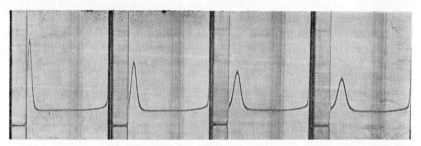

Fig. 4.6. Ultracentrifugation pattern of NADPH-adrenodoxin reductase in 50 mM potassium phosphate buffer, pH 7.4 (7.5 mg protein/ml). The photographs were taken at 8 min intervals. The rotor speed was 59,780 rpm.

of the reductase from 1 kg of bovine adrenal cortex. The basic purification steps are summarized in Table 4.6. The final preparation is homogeneous as judged by SDS-polyacrylamide gel electrophoresis and ultracentrifugation (Fig. 4.6). Recently, Sugiyama and Yamano have introduced a novel procedure for purifying the reductase by affinity chromatography using an adrenodoxin Sepharose. After chromatography on adrenodoxin Sepharose and on Sephadex G-100, the reductase is crystallized as rhombic plates.[30,31]

B. Properties

The reductase catalyzes the adrenodoxin-dependent reduction of cytochrome c in the presence of NADPH.[28] The specific activity is found to be 63 mkat/kg protein when adrenodoxin is present in excess. The reductase has a rather broad pH optimum at 7.5–9.5. The K_m value for NADPH is 1.8×10^{-6} M. That for NADH is 5.6×10^{-3} M.[32] The Michaelis constant for NADPH is about 3000 times less than that for NADH. NADP$^+$ inhibits NADPH-cytochrome c reductase activity and the K_i for NADP$^+$ has been found to be 5.3×10^{-6} M.[32] The reductase also shows adrenodoxin-dependent diaphorase activity towards 2,6-dichlorophenol-indophenol, with an activity of 9 mkat/kg protein.

However, NADH diaphorase activity is not accelerated by adrenodoxin.

The absorption spectrum of the oxidized enzyme shows absorption maxima at 378 and 450 nm and shoulders at 425 and 475 nm. Its spectrum exhibits a maximum at 272 nm in the ultraviolet region and the absorbance ratio $A_{272\ nm}/A_{450\ nm}$ is 7.3.[28,29] The midpoint potential for the reductase is estimated at -0.274 V.[32] Fully reduced enzyme can be obtained with sodium dithionite, but when NADPH is added anaerobically to the reductase, full reduction is not achieved. The NADPH-reduced enzyme shows a spectrum with low, broad long-wavelength absorbance. However, the spectrum of the reductase reduced by an NADPH-generating system consisting of a catalytic amount of NADP+, isocitrate, and isocitrate dehydrogenase, shows no long-wavelength absorbance. This suggests that the spectrum observed on reduction of the reductase with an equimolar or excess amount of NADPH is due to complex formation between NADP(H) and the reductase.[33]

The reductase is capable of forming a protein complex with adrenodoxin at a molecular ratio of 1:1.[34] Complex formation does not alter the potential of the reductase, but changes that of adrenodoxin by -40 mV. The low dissociation constants for both the oxidized and reduced forms of the complex indicate that the complex must remain associated throughout its catalytic cycle.[35] Nakamura et al.[34] have investigated the effect of various proteins on the O_2 uptake of the reductase-adrenodoxin complex, and observed that among the proteins tested, cytochrome c, histone and myoglobin are effective in stimulating the oxidase activity of the complex. The binding site of NADPH in the reductase has been examined by studies on the effects of photo-oxidation and chemical modifications of amino acid residues in the reductase.[36] The results indicate that a histidyl residue and a cysteinyl residue of the reductase are essential for the binding of NADPH by the reductase.

The enzyme flavin is liberated from the protein by heat or acid treatment. Paper chromatography of the flavin reveals a yellow spot with the same mobility as authentic FAD. The fluorescence of the enzyme flavin at 523 nm is also identical to that of authentic FAD when it is activated by light having wavelengths of 470, 375, and 275 nm. The enzyme flavin can fully activate both apo-D-amino acid oxidase and apo-salicylate hydroxylase from *Pseudomonas putida*, which are specifically reactivated by FAD.[37] The apo-enzyme is also reactivated by FAD, but not by FMN. These results indicate that FAD is the sole flavin component of the reductase, and exclude the presence of FMN as a co-component.[12,28,38]

The molecular weight of the reductase has been reported by several workers.[28,29,31,32,39] The molecular weight of the native enzyme as determined by gel filtration with Sephadex G-100 is about 54 kilodaltons, while sedimentation equilibrium experiments have yielded a molecular weight of 49.5–51 kilodaltons. Using a Sephadex G-100 column, the Stokes radius of the reductase is calculated as 3.29 nm. By substitution of this value into the Stokes-Einstein equation, a value of 6.10×10^{-7} $cm^2 \cdot sec^{-1}$ is calculated for the diffusion coefficient, $D_{20,w}$. The $s_{20,w}$ derived from the sedimentation velocity is estimated to be 3.9 S. When the reductase is treated with sodium dodecyl sulfate in the presence of 2-mercaptoethanol, the molecular weight is estimated to be 50–52 kilodaltons by SDS-polyacrylamide gel electrophoresis. The protein weight per flavin is 55 kilodaltons. These results indicate that NADPH-adrenodoxin reductase is composed of a single polypeptide chain containing one FAD. Two similar reductases to NADPH-adrenodoxin reductase, i.e. NADH-putidaredoxin reductase in camphor methylene hydroxylase from *Pseudomonas putida*[40] and NADH-rubredoxin reductase in the ω-hydroxylation system of fatty acids and hydrocarbons from *Pseudomonas oleovorance*,[41] have molecular weights of 43.5 and 55 kilodaltons, respectively. Both contain one molecule of FAD in a single chain protein. An amino acid analysis of NADPH-adrenodoxin reductase has been carried out by Chu and Kimura,[32] and has been compared with results for the above two flavoproteins. There is a great deal similarity in the amino acid compositions, especially in the high contents of hydrophobic residues. These reductases are known to contain a small amount of carbohydrate.[31,32] The formation of a stoichiometric 1:1 reductase-iron-sulfur protein complex has been reported with these reductases.[33,42] However, nonexchangeability of catalytic activities between the reductases and iron-sulfur proteins is observed among the three systems.[42,43]

4.2.4 Cytochrome P-450

A. Purification

Two species of adrenal P-450 cytochromes, catalyzing cholesterol side-chain cleavage (P-450$_{scc}$) and steroid hydroxylation (P-450$_{11\beta}$), have tended to resist purification because of ther particulate nature. Attempts to purify P-450$_{scc}$ and P-450$_{11\beta}$ have been made by a number of groups in recent years.[44–52] The method of Schleyer et al.[47] for isolating cytochrome P-450 that shows the typical absorption spectrum of the low spin form, consists of freeze drying, extraction with organic solvents, and a fractional extraction procedure with Triton N-101. The preparation obtained is active in a reconstituted hydroxylase system for the 11β-hydroxylation of deoxycorticosterone with a turnover number of 5–10 min^{-1} at 25°C, as well as in the side-chain cleavage of cholesterol (<2 min^{-1}). The specific content of cytochrome P-450 in this preparation is 1.5–2.0 nmole per mg of protein. Mitani and Horie[45] have isolated a preparation containing both high and low spin forms of cytochrome P-450 by extraction with sodium cholate, gel filtration, and ammonium sulfate fractionation. The preparation contains approximately 5 nmole heme per mg of protein. They have also obtained high spin-type cytochrome P-450 from autolyzed mitochondria by sonication, differential extraction with cholate, fractionation by ammonium sulfate, and gel filtration.[46] The specific content of heme is 4–5 nmole per mg of protein. The preparation obtained is active in side-chain cleavage reactions. The reported activity (nmole pregnenolone/30 min/nmole P-450) is 181 for 20α-hydroxycholesterol and 13 for cholesterol, respectively, at 37°C. Shikita and Hall[49] have purified P-450$_{scc}$ which is specific for the side-chain cleavage of cholesterol with a turnover number of 1.9 min^{-1} at 37°C. Their purification procedure consisted of cholate extraction, ammonium sulfate fractionation and subsequent chromatographies on DEAE-cellulose, hydroxylapatite and Bio-gel, respectively. The specific content of cytochrome P-450 was 2.9 nmole per mg of protein. Ramseyer and Harding[48] have also reported the preparation of P-450$_{scc}$ without 11β-hydroxylase activity. Jefcoate et al.[50,51] have partially separated P-450$_{11β}$ and P-450$_{scc}$ from lyophilized mitochondria by isooctane extraction and ammonium sulfate fractionation. Recently, Takemori et al.[53–55] have introduced a procedure for purifying P-450$_{scc}$ and P-450$_{11β}$ by affinity chromatography using an aniline-substituted Sepharose. This application of affinity chromatography to P-450 purification can be outlined as follows. The buffer used throughout the purification procedure consists of 50 mM potassium phosphate, 100 μM EDTA and 100 μM dithiothreitol except that, in the case of P-450$_{11β}$, 10 μM deoxycorticosterone is further added as a stabilizing agent. Mitochondrial pellets (P$_1$) prepared as in section 4.2.1 are suspended in 100 mM potassium phosphate buffer (pH 7.0). To this suspension, 100 μM dithiothreitol, sodium cholate (1 mg/1.5 mg protein), 100 μM EDTA and 10 μM deoxycorticosterone are added. After centrifugation (80,000g, 90 min), the solubilized extract is fractionated with ammonium sulfate (18 g/100 ml) and centrifuged. The P-450$_{11β}$ is mostly in the precipitated fraction and the P-450$_{scc}$, in the supernatant fraction. To the supernatant fraction, ammonium sulfate (10 g/100 ml) is further added. The precipitated P-450$_{scc}$ fraction is dissolved, dialyzed against the buffer and applied to an aniline-substituted Sepharose column (2 × 15 cm) equilibrated with the buffer. The column is washed with the buffer containing 150 mM KCl and eluted with 500 mM KCl and 0.3% sodium cholate in the buffer (pH 7.5). After the combined fractions with an absorbance ratio ($A_{393\,nm}/A_{280\,nm}$) higher than 0.25 are concentrated with ammonium sulfate, the precipitate is dissolved and dialyzed against the buffer containing 0.3% sodium cholate and 200 mM KCl. The dialyzate is applied to an aniline-substituted Sepharose column (0.7×8 cm) and eluted with the same buffer (pH 7.5). The fractions ($A_{393\,nm}/A_{280\,nm} > 0.8$) are combined and concentrated with ammonium sulfate. After dialysis against the buffer, the purified P-450$_{scc}$ obtained is stored at −80°C. In order to purify P-450$_{11\beta}$, the precipitated P-450$_{11\beta}$ fraction from the first ammonium sulfate fractionation is dissolved in the buffer (pH 7.5) containing 1% sodium cholate and 1 M KCl. After treatment with alumina C$_\gamma$ gel, the preparation is dialyzed against the buffer. Most of the P-450$_{11\beta}$ precipitates during dialysis. The precipitate is dissolved in the buffer containing 0.7% sodium cholate and 200 mM KCl. The preparation is applied to an aniline-substituted Sepharose column (1.5×3.5 cm) previously equilibrated with the same buffer. The P-450$_{11\beta}$ is eluted with the buffer (pH 7.5) containing 0.5% sodium cholate, 500 mM KCl and 0.5% Tween 20. The combined fractions ($A_{393\,nm}/A_{280\,nm} > 0.8$) are dialyzed against the buffer containing 0.3% sodium cholate and 0.3% Tween 20, and stored at 4°C. The yield is usually about 16 mg of each P-450 from 2.5 g protein of the mitochondrial particles. The specific content of heme in each purified preparation is 12 nmole per mg of protein. The turnover number (mole of product formed/min/mole heme) for P-450$_{scc}$ is 18 with cholesterol and, for P-450$_{11\beta}$, 120 and 20 at the 11β- and 18-positions, respectively, with deoxycorticosterone.

B. Properties

The absorption spectrum of the preparation isolated by Schleyer et al.[47] exhibits absorption maxima at 280, 356, 416, 541 and 570 nm, respectively, indicating the low spin-type ferric form. The EPR spectrum shows the characteristic features of the low spin form of cytochrome P-450, having signals at $g = 1.91$, 2.24 and 2.42. Ando and Horie[46] and Boyd et al.[51,56] have performed spectral analyses of P-450$_{scc}$. The spectrum of oxidized P-450$_{scc}$ exhibits absorption maxima at 393, 520 and 648 nm, typical of a high spin oxidized ferric heme protein. Addition of the product of the cholesterol side-chain cleavage reaction, pregnenolone, to such high spin state P-450$_{scc}$, results in a reversion to the low spin species. A variety of factors in the environment of P-450$_{sc}$, such as the hydrogen concentration, can convert the high spin cytochrome into the low spin form. 20α-Hydroxycholesterol causes a change in the spectrum to the low spin type. The preparation has relatively weak

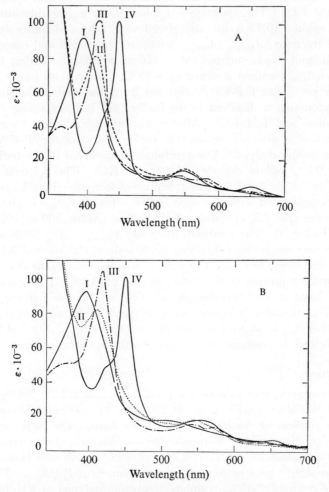

Fig. 4.7. Absorption spectra of P-450$_{SCC}$ (A) and P-450$_{11\beta}$ (B). P-450$_{SCC}$ was measured in 50 mM potassium phosphate buffer, pH 7.4, containing 100 μM EDTA, 100 μM dithiothreitol and 0.01% sodium cholate. P-450$_{11\beta}$ was measured in the same buffer system, except that 0.3% cholate and 0.3% Tween 20 were present. I, Oxidized; II, dithionite reduced; III, after treatment with adrenal electron transfer system; IV dithionite-reduced CO-complex.

low spin EPR signals (2.43, 2.52 and 1.91) and a signal at $g = 1.99$. On addition of 20α-hydroxycholesterol, the intensity of the low spin signals increases, but that of the signal at $g = 1.99$ remains unchanged.

The absorption spectra of cytochrome P-450 purified using an aniline-substituted Sepharose are illustrated in Fig. 4.7. The P-450$_{scc}$ and P-450$_{11\beta}$ preparations have very similar absorption spectra. Either purified preparation, in its ferric state, shows a high spin-type absorption spectrum with maxima at 394 and 645 nm, and a shoulder at around 540 nm. The CO-spectrum in the dithionite-reduced form gives maxima at 448 and 550 nm, respectively. The purified P-450$_{scc}$ contains endogenous cholesterol. P-450$_{11\beta}$ contains deoxycorticosterone which is added as a stabilizing agent throughout purification but does not have any significant amount of cholesterol. The steroid substrate bound to the preparation cannot be removed by extensive dialysis either before or after reduction with dithionite. However, when either P-450 is treated with the adrenodoxin-dependent electron transfer system, the absorption maxima of the preparation are shifted completely to 418, 537 and 570 nm, respectively. The substrate-free low spin form P-450$_{scc}$ reverts specifically to the high spin species on the addition of cholesterol. The substrate for 11β-hydroxylase, deoxycorticosterone, does not induce any such interconversion between low and high spin-type absorption spectra. Upon the addition of 11β-hydroxylatable steroid (deoxycorticosterone, 11-deoxycortisol or testosterone), the low spin form P-450$_{11\beta}$ spectrum is also replaced by the high spin- or substrate-bound-type. Steroids such as cholesterol, progesterone and 11β-hydroxycorticosteroid do not induce any change in the spectrum of P-450$_{11\beta}$. The cholesterol-bound high spin form P-450$_{scc}$ changes to the low spin form with a spectral shift to 416 nm on binding with 20α-hydroxycholesterol, 22R-hydroxycholesterol and pregnenolone. Deoxycorticosterone-bound high spin form P-450$_{11\beta}$ changes to the low spin species with a Soret band at 422 nm on complexation with metyrapone.[58]

The P-450$_{scc}$ preparation is specific for the side-chain cleavage reaction of cholesterol and 20α,22R-dihydroxycholesterol, and has no 11β- or 18-hydroxylase activity for deoxycorticosterone. The P-450$_{11\beta}$ catalyzes both 11β-and 18-hydroxylation of deoxycorticosterone. Deoxycorticosterone, 11-deoxycortisol, 4-androstene-3,17-dione, and possibly testosterone, are hydroxylatable substrates for P-450$_{11\beta}$ at the 11β and 19-positions. Testosterone, 11-deoxycortisol and 4-androstene-3,17-dione inhibit competitively the 11β- and 18-hydroxylation reactions of deoxycorticosterone. Metyrapone and spironolactone are also inhibitors for P-450$_{11\beta}$.[58] Rabbit antibody produced against each P-450 inhibits specifically the corresponding enzyme reaction.[59]

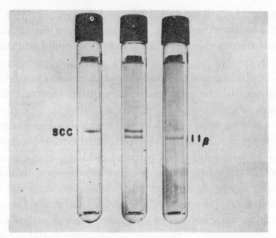

Fig. 4.8. SDS-polyacrylamide gel electrophoresis of P-450$_{SCC}$ (left), P-450$_{11\beta}$ (right), and an equimolar mixture of both (middle).

Shikita and Hall[60] have reported that P-450$_{SCC}$ is usually isolated in a form with a molecular weight of 850 kilodaltons, apparently composed of 16 subunits each of 52 kilodaltons molecular weight, and this form of P-450$_{SCC}$ can be converted to other forms having 1/2 to 1/4 of the above molecular weight. Purified preparations of P-450$_{SCC}$ and P-450$_{11\beta}$ are essentially homogeneous with $s_{20,w}$ values of 5.7 S and 3.3 S, respectively. The molecular weight of native P-450$_{SCC}$ has been calculated as 200 kilodaltons by the method of Yphantis. Reduced and carboxymethylated preparations of P-450$_{SCC}$ and P-450$_{11\beta}$ have molecular weights of 48 and 45 kilodaltons, respectively, in 6 M guanidine-HCl. As shown in Fig. 4.8, when a mixture of equal amounts of P-450$_{SCC}$ and P-450$_{11\beta}$ is used, two distinct bands are clearly separated. By comparing the mobilities of the two components with marker proteins, apparent molecular weights of 51 and 46 kilodaltons are estimated for P-450$_{SCC}$ and P-450$_{11\beta}$, respectively. P-450$_{11\beta}$ is less soluble than P-450$_{SCC}$ and is more hydrophobic, although both cytochromes are similar in their hydrophobic amino acid contents. Judging from these findings, it appears likely that P-450$_{SCC}$ and P-450$_{11\beta}$ possess many similar analytical characteristics, optical properties, and molecular size per subunit. However, they are different in their enzymatic, SDS-gel electrophoretic, hydrodynamic and immunological properties.[53-55,57-59]

4.2.5 General Discussion

The spin states of cytochrome P-450 correlate well with the optical spectra. The appearance of absorption bands at 394 and 645 nm is characteristic of the formation of a high spin form of the cytochrome, while the appearance of absorption bands at 418, 537 and 570 nm indicates the formation of a low spin form of the cytochrome. From optical absorption spectral and EPR studies it has become apparent that adrenal cytochrome P-450 exists in both a high spin and a low spin state. Jefcoate and Boyd[56] have reported that the majority of the 11β-hydroxylase activity in adrenal mitochondria is associated with a low spin form of cytochrome P-450, while a distinct high spin form of cytochrome P-450 is responsible for most of the cholesterol side-chain cleavage activity. It has been demonstrated that the elevated high spin cytochrome P-450 content of mitochondria from ether-stressed animals correlates with the increased cholesterol side-chain cleavage activity observed in these mitochondria.[51] The $P\text{-}450_{scc}$ preparation purified with an aniline-substituted Sepharose has proved extremely valuable in clarifying the correlation between cholesterol transport to the active center of cytochrome P-450 and the change from low spin to high spin cytochrome P-450.[53,57] When the original high spin-type $P\text{-}450_{scc}$ is treated with the adrenodoxin-dependent electron transfer system, the high spin-type spectrum is converted to the low spin type. The original high spin-type $P\text{-}450_{scc}$ contains endogenous cholesterol (0.6 to 1.0 mole/mole heme) and the conversion to the low spin-type spectrum is intimately associated with the appearance of the product, pregnenolone. However, this spectral change is not due to complex formation between the $P\text{-}450_{scc}$ and pregnenolone produced, since after passage through Sephadex G-25 the preparation contains no pregnenolone. On the addition of cholesterol, the low spin species is converted to the high spin species. Jefcoate and Boyd[56] have shown that addition of the product, pregnenolone, to the high spin-type P-450 results in a low spin-type spectrum. Similar phenomena are also observed with the high spin-type $P\text{-}450_{scc}$ purified with an aniline-substituted Sepharose. However, when the reaction mixture is passed through Sephadex G-25, the high spin-type spectrum again appears, suggesting that the spectral change in such a case is not due to exchange between bound cholesterol and the added pregnenolone.[57] These results indicate that the low spin-type $P\text{-}450_{scc}$ prepared by treatment with the electron donating system is the substrate-free $P\text{-}450_{scc}$. The cholesterol-bound $P\text{-}450_{scc}$ is almost identical to that of the high spin camphor complex of cytochrome P-450 obtained from *Pseudomonas putida* ($P\text{-}450_{CAM}$). $P\text{-}450_{CAM}$ is in the oxidized high spin state only when it is complexed with the substrate, camphor.[61,62] It thus appears likely that the high spin state is derived from a complex of $P\text{-}450_{scc}$ with the endogenous substrate, cholesterol.

Two enzyme activities for 11β- and 18-hydroxylations of deoxycorticosterone in the adrenal cortex have been reported by Nakamura et al.[63] to be localized in the mitochondrial fraction. Using crude mitochondrial extracts, Björkhem and Karlmar[64] first suggested that identical or at least very similar types of P-450 are involved in the 11β- and 18-hydroxylations of deoxycorticosterone. The P-450$_{11\beta}$ purified with an aniline-substituted Sepharose has provided evidence for the participation of a distinct cytochrome P-450 in the 11β- and 18-hydroxylations of deoxycorticosterone.[58] The purified P-450$_{11\beta}$ exhibits 11β- and 18-hydroxylase activity for deoxycorticosterone and 11β- and 19-hydroxylase activity for 4-androstene-3,17-dione in a reconstituted enzyme system consisting of adrenodoxin reductase, adrenodoxin and P-450. Various characteristics of these hydroxylase activities may be summarized as follows. (1) Each P-450$_{11\beta}$ preparation obtained at various purification steps exhibits a constant ratio of hydroxylase activity at the 11β- to that at the 18-position of deoxycorticosterone hydroxylation of corticosteroids. (2) The K_m values for 11β- and 18-hydroxylases of deoxycorticosterone are almost the same. (3) Steroids added to substrate-free P-450$_{11\beta}$ stabilize the 11β- and 18-hydroxylase activities to the same extent. (4) Immunoglobulin against P-450$_{11\beta}$ inhibits these activities in a similar manner. These facts suggest that only one cytochrome P-450, P-450$_{11\beta}$, is responsible for the hydroxylation of deoxycorticosterone at the 11β- and 18-positions and androgens, at the 11β- and 19-positions. Participation by other cytochromes can be excluded.

It has been proposed that the mechanism of side-chain cleavage of cholesterol involves C20- and C22- hydroxylation with cleavage of the vicinal glycol to yield pregnenolone and isocapraldehyde:[65-69]

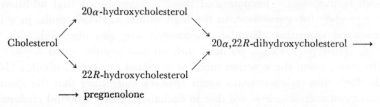

This mechanism is based on the observation that pregnenolone can be formed from either 20α-hydroxycholesterol, $22R$-hydroxycholesterol, or $20\alpha,22R$-dihydroxycholesterol by adrenal mitochondria. Recently, Shikita and Hall[70] have determined the stoichiometry of the side-chain cleavage of cholesterol, 20α-hydroxycholesterol and $20\alpha,22R$-dihydroxycholesterol with respect to NADPH, oxygen and hydrogen ion. Their results also support the above mechanism for the side-chain cleavage. On the other hand, a kinetic analysis has indicated that $20\alpha,22R$-dihydroxycholesterol may arise directly from cholesterol and not via 20α-hydroxycholesterol or 22R-hydroxycholesterol.[71,72] The question of whether the oxygen atoms of the vicinal glycol are derived from the same oxygen molecule or from separate molecules has recently been resolved by incubation of cholesterol in the presence of $^{18}O_2$ and $^{16}O_2$.[73,74] Mass spectrometric analysis indicated that the oxygen atoms of the vicinal glycol are drawn at random from the atomic pool of oxygen molecules, suggesting that the intermediate does not occur by insertion of a whole molecule of oxygen, possibly through a compound like a hydroperoxide. It has been reported that oxygen and NADPH are required for the side-chain cleavage of 20α,22R-dihydroxycholesterol.[70,75] Hall et al.[76] have demonstrated that the conversion of 20α,22R-dihydroxycholesterol to pregnenolone is inhibited by carbon monoxide and the inhibition is decreased by light of 450 nm wavelength. It would therefore appear that the cleavage of 20α,22R-dihydroxycholesterol involves a mixed function oxidation catalyzed by cytochrome P-450. However, it has been reported that the cleavage reaction in the presence of $^{18}O_2$ does not introduce this isotope into either pregnenolone or isocapraldehyde.[77] The role of oxygen in the cleavage of 20α,22R-dihydroxycholesterol thus cannot be explained satisfactorily at present.

Sih[78,79] has proposed that NADPH exerts a dual function in the steroid hydroxylation system of adrenal mitochondria. It maintains the auto-oxidizable cytochrome P-450 in the reduced state via the electron transfer system, a course which does not enter into the stoichiometry of the hydroxylation reaction. Second, NADPH is directly involved in the steroid hydroxylation reaction to generate "active oxygen" $(Fe^{2+}-O-OH)^-$, a process which is steroid-dependent. On the other hand, Huang and Kimura[80] have demonstrated that steroid hydroxylation occurs exclusively without NADPH if reduced adrenodoxin is supplied as a specific electron donor.

The cholesterol side-chain cleavage reaction also occurs in mitochondria of the corpus luteum,[81] testis[82] and placenta.[83] The terminal enzyme that catalyzes the conversion of cholesterol to pregnenolone is cytochrome P-450.[84–86] The electron transfer system of human placental mitochondria has been shown to consist of an NADPH-specific flavoprotein, an iron-sulfur protein and a P-450 cytochrome.[87] Electron transfer components similar to those of adrenal cortex are also found to occur in the corpus luteum.[85] In chick kidney, a mitochondrial enzyme system catalyzes the 1α-hydroxylation of 25-hydroxycholecalciferol to 1,25-dihydroxycholecalciferol. Ghazarian et al.[88] have demonstrated that chick kidney mitochondria contain cytochrome P-450. Studies on the solubilized enzyme system revealed that cytochrome P-450 functions as a 1α-hydroxylase when supplemented with NADPH and NADPH-adrenodoxin reductase and adrenodoxin. These findings suggest that a flavoprotein, an iron-sulfur protein and a P-450 cytochrome may be general components of mitochondrial electron transfer steroid hydroxylase systems.

References

Section 4.2
1. M. L. Sweat and M. D. Lipscomb, *J. Am. Chem. Soc.*, **77**, 5185 (1955).
2. J. K. Grant and A. C. Brownie, *Biochim. Biophys. Acta*, **18**, 433 (1955).
3. M. Hayano, M. C. Lindberg, R. I. Dorfman, J. E. H. Hancock and W. V. E. Doering, *Arch. Biochem. Biophys.*, **54**, 529 (1955).
4. M. L. Sweat, R. A. Aldrich, C. H. Debruin, W. L. Fowlks, L. R. Heiselt and H. S. Mason, *Fed. Proc.*, **15**, 367 (1956).
5. B. W. Harding, S. H. Wong, and D. H. Nelson, *Biochim. Biophys. Acta*, **92**, 415 (1964).
6. B. W. Harding and D. H. Nelson, *J. Biol. Chem.*, **241**, 2212 (1966).
7. T. Kinoshita, S. Horie, N. Shimazono and T. Yohro, *J. Biochem. (Tokyo)*, **60**, 391 (1966).
8. S. Horie, T. Kinoshita and N. Shimazono, *ibid.*, **60**, 660 (1966).
9. D. Y. Cooper, B. Novack, O. Foroff, A. Slade, E. Saunders, S. Narasimhulu and O. Rosenthal, *Fed. Proc.*, **26**, 341 (1967).
10. Y. Nakamura, H. Otsuka and B. Tamaoki, *Biochim. Biophys. Acta*, **122**, 34 (1966).
11. T. Omura, E. Sanders, R. W. Estabrook, D. Y. Cooper and O. Rosenthal, *Arch. Biochem. Biophys.*, **117**, 660 (1966).
12. T. Kimura and K. Suzuki, *J. Biol. Chem.*, **242**, 485 (1967).
13. I. D. K. Halkerston, J. Eichhorn, and O. Hechter, *ibid.*, **236**, 374 (1961).
14. E. R. Simpson and G. S. Boyd, *Eur. J. Biochem.*, **2**, 275 (1967).
15. E. R. Simpson, and G. S. Boyd, *Biochem. Biophys. Res. Commun.*, **28**, 945 (1967).
16. A. Y. H. Lu and M. J. Coon, *J. Biol. Chem.*, **243**, 1331 (1968).
17. K. Suhara, S. Takemori and M. Katagiri, *Biochem. Biophys. Acta*, **263**, 272 (1972).
18. T. Kimura, *Structure and Bonding* (ed. C. K. Jørgensen, J. B. Neilands, R. S. Nybolm, D. Reinen and R. J. P. Williams) vol. 5, p. 1, Springer-Verlag, (1968).
19. T. Kimura and J. J. Huang, *Arch. Biochem. Biophys.*, **137**, 357 (1970).
20. K. Suhara, K. Kanayama, S. Takemori and M. Katagiri, *Biochem. Biophys. Acta*, **336**, 309 (1974).
21. R. W. Estabrook, K. Suzuki, J. I. Mason, J. Baron, W. E. Taylor, E. R. Simpson, J. Purvis and J. McCarthy, *Iron-sulfur Proteins* (ed. W. Lovenberg), vol. 1, p. 193, Academic Press (1973).
22. J. K. Fogo and M. Popowsky, *Anal. Chem.*, **21**, 732 (1949).
23. K. Suhara, S. Takemori, M. Katagiri, K. Wada, O. Kobayashi and H. Matsubara *Anal. Biochem.* **68**, 632 (1975).
24. M. Tanaka, M. Haniu and K. T. Yasunobu, *Biochem. Biophys. Res. Commun.* **39**, 1182 (1970)
25. M. Tanaka, M. Haniu, K. T. Yasunobu and T. Kimura, *J. Biol. Chem.*, **248**, 1141 (1973).
26. H. Watari and T. Kimura, *Biochem. Biophys. Res. Commun.*, **24**, 106 (1966).
27. H. P. Misra and I. Fridovich, *J. Biol. Chem.*, **246**, 6886 (1971).
28. K. Suhara, Y. Ikeda, S. Takemori and M. Katagiri, *FEBS Lett.*, **28**, 45 (1972).

29. K. Suhara, Y. Ikeda, S. Takemori and M. Katagiri, *Flavins and Flavoproteins* (ed. T. P. Singer), p. 655, Elsevier (1976).
30. T. Sugiyama and T. Yamano, *FEBS Lett.*, **52**, 145 (1975).
31. A. Hiwatashi, Y. Ichikawa, N. Maruya, T. Yamano and K. Aki, *Biochemistry*, **15**, 3082 (1976).
32. J-W. Chu and T. Kimura, *J. Biol. Chem.*, **248**, 2089 (1973).
33. J. D. Lambeth and H. Kamin, *Flavins and Flavoproteins* (ed. T. P. Singer), p. 647, Elsevier (1976).
34. S. Nakamura, T. Kimura and J-W. Chu, *FEBS Lett.*, **25**, 249 (1972).
35. J. D. Lambeth, D. R. McCaslin and H. Kamin, *J. Biol. Chem.*, **251**, 7545 (1976).
36. A. Hiwatashi, Y. Ichikawa, T. Yamano and N. Maruya, *Biochemistry* **15**, 3091 (1976).
37. S. Yamamoto, M. Katagiri, H. Maeno and O. Hayaishi, *J. Biol. Chem.*, **240**, 3408 (1965).
38. T. Omura, E. Sanders, D. Y. Cooper and R. W. Estabrook, *Methods in Enzymology* (ed. S. P. Colowick and N. O. Kaplan), vol. 10, p. 362, Academic Press (1967).
39. R. P. Foster and L. D. Wilson, *Biochemistry*, **14**, 1477 (1975).
40. R. L. Tsai, I. C. Gunsalus and K. Dus, *Biochem. Biophys. Res. Commun.*, **45**, 1300 (1971).
41. T. Ueda, E. T. Lode and M. J. Coon, *J. Biol. Chem.* **247**, 2109 (1972).
42. T. Ueda and M. J. Coon, *ibid.*, **247** 5010 (1972).
43. T. Kimura and H. Ohno, *J. Biochem. (Tokyo)*, **63**, 716 (1968).
44. T. Kinoshita, *Tokyo J. Med. Sci.* (Japanese), **75**, 202 (1967).
45. F. Mitani and S. Horie, *J. Biochem. (Tokyo)*, **65**, 269 (1969).
46. N. Ando and S. Horie, *ibid.*, **72**, 583 (1972).
47. H. Schleyer, D. Y. Cooper and O. Rosenthal, *J. Biol. Chem.*, **247**, 6103 (1972).
48. J. Ramseyer and B. W. Harding, *Biochim. Biophys. Acta*, **315**, 306 (1973).
49. M. Shikita and P. H. Hall, *J. Biol. Chem.*, **248**, 5598 (1973).
50. C. R. Jefcoate, R. Hume and G. S. Boyd, *FEBS Lett.*, **9**, 41 (1970).
51. G. S. Boyd, A. C. Brownie, C. R. Jefcoate and E. R. Simpson, *Biological Hydroxylation Mechanisms* (ed. G. S. Boyd and R. M. S. Smellie), p. 207, Academic Press (1972).
52. H-P. Wang and T. Kimura, *J. Biol. Chem.*, **251**, 6068 (1976).
53. S. Takemori, K. Suhara, S. Hashimoto, M. Hashimoto, H. Sato, T. Gomi and M. Katagiri, *Biochem. Biophys. Res. Commun.*, **63**, 588 (1975).
54. S. Takemori, H. Sato, T. Gomi, K. Suhara and M. Katagiri, *ibid.*, **67**, 1151 (1975).
55. M. Katagiri, S. Takemori, E. Itagaki, K. Suhara, T. Gomi and H. Sato, *Iron and Copper Proteins* (ed. K. T. Yasunobu, H. F. Mower and O. Hayaishi), p. 281, Plenum (1976).
56. C. R. Jefcoate and C. S. Boyd, *FEBS Lett.*, **12**, 279 (1971).
57. S. Takemori, T. Gomi, H. Sato, K. Suhara, E. Itagaki and M. Katagiri, *unpublished data*.
58. H. Sato, K. Suhara, E. Itagaki, S. Takemori and M. Katagiri, *unpublished data*.
59. K. Suhara, H. Sato, E. Itagaki, S. Takemori and M. Katagiri, *unpublished data*.
60. M. Shikita and P. F. Hall, *J. Biol. Chem.*, **248**, 5605 (1973).
61. M. Katagiri, B. N. Ganguli and I. C. Gunsalus, *ibid.*, **243**, 3543 (1968).
62. C-A. Yu, I. C. Gunsalus, M. Katagiri, K. Suhara and S. Takemori, *ibid.*, **249**, 94 (1974).
63. Y. Nakamura, H. Otsuka and B. Tamaoki, *Biochim. Biophys. Acta*, **96**, 339 (1965).
64. I. Björkhem and K-E. Karlmar, *Eur. J. Biochem.*, **51**, 145 (1975).
65. G. Constantopoulos and T. T. Tschen, *J. Biol. Chem.*, **236**, 65 (1961).
66. K. Shimizu, M. Hayano, M. Gut and R. I. Dorfman, *ibid.*, **236**, 695 (1961).
67. K. Shimizu, M. Gut and R. I. Dorfman, *ibid.*, **237**, 699 (1962).
68. G. Constantopoulos, A. Carpenter, P. S. Satoh and T. T. Tschen, *Biochemistry*, **5**, 1650 (1966).

69. A. C. Chaudhuri, Y. Harada, K. Shimizu M. Gut and R. I. Dorfman, *J. Biol. Chem.*, **237**, 703 (1962).
70. M. Shikita and P. F. Hall, *Proc. Natl. Acad. Sci. U.S.A.*, **71**, 1441 (1974).
71. S. Burstein, H. L. Kimball and M. Gut, *Steroids*, **15**, 809 (1970).
72. S. Burstein and M. Gut, *Rec. Progr. Hormone Res.*, **27**, 303 (1971).
73. S. Burstein, B. S. Middleditch and M. Gut *Biochem. Biophys. Res. Commun.*, **61** 692 (1974).
74. S. Burstein, B. S. Middleditch and M. Gut, *J. Biol. Chem.* **250**, 9028 (1975).
75. K. Shimizu, *Arch. Biochem. Biophys.*, **125**, 1016 (1968).
76. P. F. Hall, J. L. Lewes and E. D. Lipson, *J. Biol. Chem.*, **250**, 2283 (1975).
77. C. Takemoto, H. Nakano, H. Sato and B. Tamaoki, *Biochim. Biophys. Acta*, **152**, 749 (1968).
78. C. J. Sih, Y. Y. Tsong and B. Stein, *J. Am. Chem. Soc.*, **90**, 5300 (1968).
79. C. J. Sih, *Science*, **163**, 1297 (1969).
80. J. J. Huang and T. Kimura, *Biochem. Biophys. Res. Commun.*, **41**, 737 (1970).
81. N. Yago, R. I. Dorfman and E. Forchielli, *J. Biochem. (Tokyo)*, **62**, 345 (1967).
82. R. I. Dorfman, K. M. J. Menon, D. C. Sharma and E. Forchielli, *Biogenesis and Action of Steroid Hormones* (ed. R. I. Dorfman K. Yamazaki and M. Dorfman), p. 398, Geron-X, Inc., Los Angeles (1968).
83. K. J. Ryan, R. Meigs and L. Petro *Am. J. Obstet. Gynecol.*, **96**, 676 (1966).
84. T. Yohro and S. Horie, *J. Biochem. (Tokyo)*, **61**, 515 (1967).
85. E. N. McIntosh, F. Mitani, V. I. Uzgiris, C. Alonso and H. A. Salhanick. *Ann. N. Y. Acad. Sci.*, **212**, 392 (1973).
86. R. A. Meigs and K. J. Ryan, *Biochim. Biophys. Acta*, **165**, 476 (1968).
87. J. I. Mason, and G. S. Boyd, *Eur. J. Biochem.*, **21**, 308 (1971).
88. J. G. Ghazarian, C. R. Jefcoate, J. C. Knutson, W. H. Orme-Johnson and H. F. Deluca, *J. Biol. Chem.*, **249**, 3026 (1974).

4.3 Bacterial Systems

The availability of soluble cytochrome P-450 from bacteria has permitted the purification of this cytochrome to homogeneity, providing the direct and most convincing evidence regarding its spectral and molecular properties, enzymatic function, and the mechanism of the hydroxylation catalyzed by cytochrome P-450-containing systems.

At least four microbial P-450 cytochromes are known to occur as readily soluble proteins. These cytochromes are camphor 5-*exo* monooxygenase from *Pseudomonas putida*[1,2] (P-450$_{CAM}$), cytochrome P-450 from cholesterol-grown *Corynebacterium* sp. (P-450$_{CB}$),[3] cytochrome P-450 from *Rhizobium japonicum*,[4] and chloroperoxidase from *Caldariomyces*.[5] The last enzyme is an extracellular protein which is released into the medium during the growth of the cells under normal culture conditions. *n*-Alkane hydroxylating systems from *Corynebacterium* sp.[6,7] and *Candida tropicalis*[8] represent two very similar enzyme systems. The solubilized extracts have, respectively, been separated into two protein fractions, one of which shows the spectral characteristics of a P-450 cytochrome.

Both fractions are required for n-alkane hydroxylation in the presence of a reduced pyridine nucleotide and molecular oxygen. Unfortunately, however, these n-alkane hydroxylating cytochrome P-450 components, referred to as P-450$_\omega$ in this section, are "particulate". Four* among the six P-450 cytochromes mentioned, viz. P-450$_{CAM}$, P-450$_{CB}$, and the two P-450$_\omega$'s, are induced proteins in which the inducers are D-camphor, cholesterol, n-octane and tetradecane, respectively.

In view of the fact that animal P-450 cytochromes are not isolated in a readily soluble form without resorting to the use of detergents or other solubilizing agents, it is clear that the use of bacterial P-450 cytochromes is advantageous in studying both the molecular and functional properties of cytochrome P-450-containing hydroxylase systems. Indeed, many of our present general and fundamental concepts on the spectral, physical, chemical and functional properties of cytochrome P-450 are based on the evidence provided by studies on purified P-450$_{CAM}$.[11]

The main advantages related to the use of bacterial cytochrome P-450 may be summarized as follows. (1) In contrast to the P-450 in mammalian organelles, some of the known microbial P-450 cytochromes are readily soluble proteins. (2) Once a soluble cytochrome has been found and the cultural and preparative conditions established, large quantities of pure preparation are readily available. (3) Some bacterial P-450 cytochromes are, similarly to those of liver microsomal origin, inducible proteins. The presence of a specific inducer is absolutely required for production; in other words, the availability of a variety of specific P-450 cytochromes may be expected in bacteria. (4) The turnover is high. Compared with the catalytic activities of cytochrome P-450 in animal tissues for various hydroxylatable substrates, the catalytic activity of P-450$_{CAM}$ for camphor hydroxylation is two or three orders of magnitude higher. This facilitates the use of bacterial systems for kinetic studies on the reaction mechanisms of cytochrome P-450-catalyzed hydroxylations. (5) There is higher clear-cut substrate specificity, which is closely related to the physiological role of the cytochrome P-450 in cells.

* Recently, Anders et al.[9] have presented evidence that a steroid 15β-hydroxylase system of *Bacillus megaterium* resembles the camphor 5-*exo* hydroxylase system of *Ps. putida*. The 15β-hydroxylase system is composed of an NADPH-specific flavoprotein (megaredoxin reductase), an iron-sulfur protein (megaredoxin) and a P-450 cytochrome (P-450$_{meg}$). The purified preparation of P-450$_{meg}$ sediments as a homogeneous zone on sucrose gradients, and contains one mole of heme per mole of P-450$_{meg}$ with a molecular weight of 40,000. The megaredoxin reductase is characterized as an FMN-containing flavoprotein.

A cytochrome P-450-containing system has also been identified in the O-demethylase system of *Nocardia* NH$_1$.[10] The P-450 cytochrome (P-450$_{npd}$) has been extensively purified and shown to be homogeneous. The molecular weight was reported to be 42,000 to 45,000.

4.3.1 Catalytic Components of the Camphor-Methylene Hydroxylating System (Camphor 5-*exo* Hydroxylase)

The camphor 5-*exo* hydroxylase system is formed by *Pseudomonas putida* in response to a need to degrade camphor for providing energy and supplying carbon. It is composed of three protein components: a flavoprotein (NADH-putidaredoxin reductase),[1,12] an iron-sulfur protein (putidaredoxin),[13] and a hemoprotein (cytochrome P-450$_{CAM}$).[1,2] All three components are readily soluble and each protein consists of a single polypeptide chain carrying one molecule of prosthetic group. The enzyme system catalyzes a methylene hydroxylation of a keto-terpene, i.e. the conversion of D-camphor to its 5-*exo* alcohol.[14]

The spectrophotometric properties of the cytochrome P-450 component are basically consistent with those of mammalian cytochrome P-450. The function of P-450$_{CAM}$ in the overall hydroxylation system appears to be as the terminal oxidase, coupled with the hydroxylation of the substrate. It catalyzes the hydroxylation of D- or L-camphor, or D- or L-camphor 1,2-lactone to their respective 5-*exo* alcohols in the presence of NADH, molecular oxygen and the two other protein components. In the hydroxylation reaction, P-450$_{CAM}$ has a turnover of 17 sec^{-1}, which is orders of magnitude greater than the values reported for liver microsomal and adrenocortical mitochondrial cytochrome P-450-catalyzed hydroxylation reactions.

P-450$_{CAM}$ was the first P-450 cytochrome to be purified and crystallized.[2,15] It was discovered in 1968 by Katagiri, Ganguli and Gunsalus[16] in camphor-grown *Pseudomonas putida*, and has been shown to contain one molecule of protoheme per protein of molecular weight 46,000.[1,2] It consists of a single peptide chain with threonine on the amino terminus and valine on the carboxy terminus.[17,18]

P-450$_{CAM}$ is readily extracted from the thawed frozen cells suspended in 50 mM Tris-Cl buffer (pH 7.4) containing 10 mM mercaptoethanol, either by sonic disruption or by autolysis at 10°C for 1 h. From these crude extracts, P-450$_{CAM}$ has been purified by two procedures, both in the presence of camphor as a stabilizer:[2] (1) by DEAE-cellulose chromatography, ammonium sulfate fractionation, Sephadex G-75 filtration, and calcium phosphate (hydroxylapatite) chromatography; and (2) by Bio-gel P150 and DEAE-Sephadex chromatography instead of Sephadex G-75 and calcium phosphate chromatography. The first procedure is favored for large-scale preparations and for the simultaneous isolation of the two other components of the 5-*exo*-hydroxylase system, i.e. the ferredoxin and the NADH-ferredoxin reductase. The second procedure is preferred when only P-450$_{CAM}$ is required. The P-450$_{CAM}$ preparations from each of these purification procedures have been found to be homogeneous when examined for their electrophoretic mobility and hydrodynamic properties. The purified P-450$_{CAM}$ showed typical absolute and difference spectra in combination with substrate in the oxidized state, and with carbon monoxide in the reduced state. Thus, in its spectrophotometric properties, P-450$_{CAM}$ does not differ, or differs only slightly, from animal cytochrome P-450. The absorption spectrum of the oxidized form of P-450$_{CAM}$ shows an α-band at 571 nm and a β-band at 538 nm, both with a millimolar extinction coefficient (ε_{mM}) of approximately 10. The Soret band is at 417 nm, with $\varepsilon_{mM}=105$; a shoulder or δ-band is present at 365 nm, with $\varepsilon_{mM}=36$. Absorption at 280 nm, $\varepsilon_{mM}=60$, is observed, with a shoulder at 290 nm. Reduction of P-450$_{CAM}$ with sodium dithionite was found to shift the Soret band to 411 nm, $\varepsilon_{mM}=71$, and the α- and β-bands fused into a single absorption at 540 nm, $\varepsilon_{mM}=14$. The intensities of these absorption bands were essentially constant at neutral pH values in both the oxidized and reduced forms of P-450$_{CAM}$. At pH values below 5.5 or above 8.8, however, the δ-band of the oxidized form increased and the Soret absorption decreased drastically, while the 280 nm band remained unchanged. The decrease in the Soret absorption was directly proportional to the absolute amount of active P-450$_{CAM}$ in the sample, i.e. proportional to the degree of denaturation. On addition of camphor, the Soret band shifted to 391 nm and decreased in intensity by approximately 15%, i.e. to $\varepsilon_{mM}=91$, with concomitant appearance of α-band at 646 nm, $\varepsilon_{mM}=4.3$. This has been attributed to high spin charge transfer, as indicated by EPR measurements.[19] The addition of substrate to the reduced form of P-450$_{CAM}$ resulted in a 3-nm blue shift of the Soret band, without change in its intensity. A blue shift of about 30 nm in the Soret absorption of purified P-450$_{CAM}$ upon addition of camphor facilitates first measurements of the substrate binding. P-450$_{CAM}$ was titrated with D-camphor and an estimation of the stoichiometry was made. The initial slope of the titration curve intercepted the maximum value at 1 mole D-camphor per mole of P-450$_{CAM}$. The K_d estimated for D-camphor in 50 mM phosphate buffer from these values was about 3 μM for pH 7 to 9. Ferrous (reduced) P-450$_{CAM}$ reacts with carbon monoxide to form a complex with a red spectral shift in both the free and substrate complex forms. The Soret maximum is at 446 nm, $\varepsilon_{mM}=106$, the α- and β-bands are fused at 550 nm, $\varepsilon_{mM}=12$.

P-450$_{CAM}$ is stabilized by the addition of camphor, during both its purification and storage. Over a 3-month period at $-20°C$, camphor-free preparations showed gradual deterioration; about 30% of the activity was lost at 2 mg/ml in pH 7 phosphate buffer during a storage period of 1 week at −20°C. Purified P-450$_{CAM}$ is very stable at neutral pH values, when substrate is present.

The P-450$_{CAM}$-substrate complex was stoichiometrically reduced with NADH. One-half equivalent of NADH reduces the heme iron entirely to the ferrous substrate form. On exposure to air, the ferrous complex is converted to the oxygenated intermediate with a Soret absorption at 418 nm. This discovery was first reported in 1970 by Gunsalus' group[20] and then later by Ishimura et al.[21] The oxygenated species is converted to product and oxidized P-450$_{CAM}$ when an additional electron is furnished via putidaredoxin.

Ps. putida ferredoxin has been isolated and purified by Gunsalus' group[13] by DEAE-cellulose chromatography and Bio-gel P60 filtration. The ferredoxin contains 106 amino acids arranged in a single peptide chain. The complete amino acid sequence is known, and the homology to adrenodoxin has been pointed out by Tanaka et al.[22] The molecular weight including two atoms each of iron and "labile" sulfur is 11,594. The putaredoxin shows an absorption spectrum similar to those of other ferredoxins such as adrenal and testis ferredoxins. The spectrum of the oxidized form exhibits absorption bands at 280, 325, 415, 455 and 545 nm, with millimolar extinction coefficients (ε_{mM}) of 22.5, 15.0, 10.0, 9.6 and 5.0, respectively.

The NADH-putaredoxin reductase has also been separated as an electron transfer component in the camphor 5-*exo* hydroxylase system. The reductase shows a typical flavoprotein absorption spectrum and contains one molecule of FAD in a single chain protein with a molecular weight of 43,500. These findings suggest that NADH-putaredoxin reductase is very similar to the corresponding bovine adrenocortical mitochondrial enzyme, NADPH-adrenodoxin reductase, in its molecular properties.

4.3.2 Function of P-450$_{CAM}$

Mammalian cytochrome P-450 has been recognized to participate as a component of hydroxylase systems, although much still remain to be elucidated. Studies with the camphor hydroxylating system of *Ps. putida*, which involved the isolation and purification of the components, established the general concept of cytochrome P-450 as an enzyme; P-450$_{CAM}$ is camphor 5-*exo* monooxygenase. The hydroxylation process can be followed from the rate of NADH oxidation at 340 nm in the presence of NADH, molecular oxygen and all the three protein components;

NADH-putaredoxin reductase, putaredoxin, and P-450$_{CAM}$.[1,16] The addition of substrate causes an immediate increase in NADH oxidation and leads to the stoichiometric formation of product according to the monooxygenase reaction. The turnover number for entire product formation in an enzyme system with the cytochrome P-450 limiting is about 1000. This value coincides with the value of about 1000 for a typical and more primitive reduced pyridine nucleotide-dependent monooxygenase reaction, the salicylate hydroxylase reaction.[23]

The initial reaction in the process consists of extremely rapid (7000 sec^{-1}) binding of the substrate to P-450$_{CAM}$ with transition of the latter from the low spin to the high spin form.[24] This change is observable as a complete spectral shift of the cytochrome in the Soret region. Once the cytochrome has bound to the substrate, it can undergo the first one-electron reduction through the enzymatic system, i.e. NADH, NADH-putaredoxin reductase and putaredoxin. The third step is considered to consist of an interaction of the reduced P-450$_{CAM}$-substrate complex with molecular oxygen, giving rise to a P-450$_{CAM}$-substrate-oxygen complex with a Soret absorption at 418 nm. After the second one-electron reduction via the putaredoxin system, the complex dissociates into the oxidized P-450$_{CAM}$, hydroxylated product and water.[24]

4.3.3 Cytochrome P-450-dependent n-Alkane Methyl Hydroxylation (ω-Hydroxylation)

The oxidation of n-alkanes has long been known to be readily accomplished by microbial systems and to proceed via a terminal or an ω-hydroxylation reaction which forms the corresponding primary alcohol. More recent studies on the oxidation of n-alkane by cell-free extracts of *Pseudomonas oleovorans* by Peterson *et al.*[25,26] have revealed that the enzyme system consists of three protein components, i.e. an NADH-rubredoxin reductase, a rubredoxin, and an ω-hydroxylase. The latter enzyme has been shown to be a cyanide-sensitive non-heme iron protein,[27] and cytochrome P-450 is apparently not present in the system. In contrast to the hydroxylase from *Pseudomonas*, the n-alkane hydroxylating system of liver microsomes does contain a P-450 cytochrome.[28]

Cardini and Jurtshuk[6] have reported that the n-octane oxidizing cell-free system of *Corynebacterium* sp. (strain 7ElC) is carbon monoxide-sensitive and does contain a P-450 cytochrome. Cells of *Corynebacterium* grown in a mineral salts medium containing n-octane as the sole source of carbon and energy, were disrupted by sonic oscillation. The 144,000g supernatant was separated into two fractions; a cytochrome P-450 fraction (25 to 40% saturation of ammonium sulfate) and a second fraction (60 to 100% ammonium sulfate saturation). Both fractions are required for the formation of hydroxylated product in the presence of NADH and oxygen. The presence of an iron-sulfur protein in the extract has not been recognized. The cytochrome P-450 appeared to be particulate, or is at least associated with lipids. The content of the cytochrome in the reported preparation was calculated to be a few percent when the minimum molecular weight per P-450-heme was assumed to be about 50,000. The fraction is essentially free of other hemoproteins but may contain small amounts of cytochrome P-420. Studies with inhibitors have shown that the overall hydroxylation is inhibited by carbon monoxide, but is insensitive to cyanide and azide. Comparative studies with enzyme obtained from n-octane-grown cells and acetate-grown cells have indicated that P-450$_\omega$ is an inducible protein.

The existence of a similar P-450$_\omega$ cytochrome with an n-alkane hydroxylating function in cell-free systems of *Candida tropicalis*, grown on tetradecane, has been reported.[8] The cells were suspended in a buffer containing a high concentration of glycerol and the suspension was then passed through a French pressure cell. The 27,000g supernatant was separated with ammonium sulfate into two fractions; (A) at 0 to 40% saturation, and (B) at 45 to 80% saturation. Both fractions were required for maximal activity, and one fraction (A) was found to contain a P-450 cytochrome. In contrast to the *Corynebacterium* system, however, NADPH is specifically required for the ω-hydroxylation and cannot be replaced by NADH.

4.3.4 Other Bacterial P-450-type Cytochromes

Historically speaking, the family of P-450 cytochromes obtained from *Rhizobium japonicum* by Appleby[29] represents the first bacterial P-450 cytochromes to be discovered. The *Rhizobium* P-450 cytochromes are released into solution by gentle disruption of the whole bacteria and may be purified without use of detergents. A minimum of three cytochromes have been separated by DEAE-cellulose chromatography, and they are called respectively P-450a, P-450b and P-450c.[4] Optical and EPR studies on the oxidized P-450 components have indicated that P-450a is probably a pure low spin hemoprotein similar to the typical substrate-free oxidized form of cytochrome P-450, that P-450b, with an absorption Soret peak at 396 nm and a well-defined peak at 646 nm, contains a large proportion of high spin form, and that P-450c is a mixed spin hemoprotein. However, few ligand-binding experiments have so far been performed and no evidence indicating the enzymatic nature of the cytochromes has been obtained.

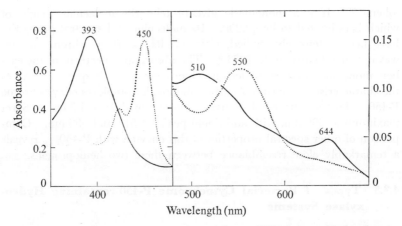

Fig. 4.9. Absorption spectra of partially purified cholesterol-induced cytochrome P-450 from *Corynebacterim*. The protein concentration was 1.6 mg/ml in 50 mM potassium phosphate buffer, pH 7.4, in the presence of 10 mM β-mercaptoethanol and 100 μM EDTA. The solid curve represents the absorption spectrum of the native form, and the dotted curve the difference between the absorbance of the $Na_2S_2O_4$-reduced form and CO-complexed form after gassing with CO for 20 min.

Recently, a soluble P-450 cytochrome has been obtained in our laboratory from *Corynebacterium* sp. grown in a mineral salts medium containing cholesterol as the sole carbon and energy source.[3] The cell-free extract was obtained from frozen cells by disruption in a sonic disintegrator or by grinding in a mortar with aluminum oxide. The cytochrome, P-450$_{CB}$, was partially purified from the extract by a procedure employing DEAE-cellulose and Sephadex G-100 chromatography. The molecular weight on a Sephadex base is about 50,000, and the current purity of the partially purified cytochrome P-450 preparation is about 40%. Protoheme has been shown to be the prosthetic group. Judging from the absorption spectra of native, carbon monoxide-complexed and aniline-complexed preparations, the cytochrome component appears to be a typical cytochrome P-450 type hemoprotein. The native absolute spectrum exhibits a Soret peak at 393 nm, with additional peaks at 510 and 645 nm. The spectrum of the carbon monoxide-complexed form shows absorption bands at 450 and 550 nm (see Fig. 4.9). No coexistent iron-sulfur protein was detected in crude sonic extracts of the cholesterol-induced *Corynebacterium* cells.

It is interesting to note that the hemoprotein, chloroperoxidase from *Caldariomyces fumago* is a P-450 cytochrome,[5] although its function is not as a hydroxylase. This chloroperoxidase was isolated in crystalline form early in 1966 and the prosthetic group has been identified as protoheme.[30]) It is a monomeric glycoprotein, the molecular weight of which is estimated to be 42,000. Its molecular and spectral properties have been extensively studied, although its P-450 cytochrome nature was not investigated until 1973.[5]) The absorption spectrum of the carbon monoxide complex of ferrous chloroperoxidase is quite similar to the characteristic spectrum of carbon monoxide complexes of cytochrome P-450. The spectrum of native chloroperoxidase exhibits a Soret maximum at 396 nm, with additional peaks at 515 and 650 nm. Comparison of other spectral properties of this enzyme with P-450$_{CAM}$ reveals a remarkably close resemblance between these two hemoproteins.

4.3.5 Types of Bacterial Cytochrome P-450-containing Hydroxylase Systems

Basically, two types of cytochrome P-450-containing hydroxylase systems are recognized in mammalian tissues, as exemplified by adrenocortical mitochondrial steroid hydroxylase systems and liver microsomal cytochrome P-450-containing xenobiotic hydroxylating systems. These two types will be referred to here as the mitochondrial type and the microsomal type, respectively. A similar binary classification also appears to be applicable to microbial systems.

The adrenocortical mitochondrial steroid hydroxylase system has been separated into three fractions; NADPH-adrenodoxin reductase, adrenodoxin, and cytochrome P-450 (*cf.* section 4.2). Similar systems are also obtained from mitochondria of other tissues. The camphor 5-*exo* hydroxylase system obtained from camphor-induced *Ps. putida* has also been separated into three fractions. A comparison of its molecular and spectral properties reveals a resemblance between the respective adrenal and *Pseudomonas* components. The major reaction catalyzed by both systems appears to be the methylene hydroxylation of keto-terpenoid compounds, viz. steroid hormones and camphor. NADPH appears to be a specific electron donor in mammalian systems, while in microbial systems, NADH is a specific donor.

The microsomal-type cytochrome P-450 system does not require iron-sulfur protein as a functional component, and the microbial cytochrome P-450 systems obtained from *n*-alkane-induced cells may also fall into this category. The enzymatic function of *n*-alkane ω-hydroxylation indicates a similarity between *Corynebacterium* and *Candida* P-450$_\omega$ and liver microsomal ω-hydroxylating cytochrome P-450. Again, NADPH is a specific electron donor in the mammalian system. However, in the microbial systems, NADH is required in *Corynebacterium*, and NADPH in *Candida*.

In contrast to the clear differences in their state (particulate or soluble) and origin (microsomal, mitochondrial or microbial), all P-450 cytochromes show the same unique spectral characteristics. The similarities described here suggest possible common structural and functional relations among all P-450 cytochromes. In this context, soluble microbial P-450 cytochromes may serve as prototypes for membrane bound P-450 cytochromes of mammalian origin.

REFERENCES

Section 4.3
1. M. Katagiri, B. N. Ganguli and I. C. Gunsalus, *J. Biol. Chem.*, **243**, 3543 (1968).
2. C-A. Yu, I. C. Gunsalus, M. Katagiri, K. Suhara and S. Takemori, *ibid.*, **249**, 94 (1974).
3. A. Yamamoto, S. Takemori and M. Katagiri, *Seikagaku (Japanese)*, **46**, 534 (1974)
4. C. A. Appleby and R. M. Daniel, *Oxidases and Related Redox Systems* (ed. T.E. King, H. S. Mason and M. Morrison), p. 515, University Park Press (1973).
5. P. F. Hollenberg and L. P. Hager, *J. Biol. Chem.*, **248**, 2630 (1973).
6. G. Cardini and P. Jurtshuk *ibid.*, **243**, 6070 (1968).
7. G. Cardini and P. Jurtshuk, *ibid.*, **245**, 2789 (1970).
8. J-M. Lebeault, E. T. Lode and M. J. Coon, *Biochem. Biophys. Res. Commun.*, **42**, 413 (1971).
9. B. Anders, J-Å. Gustafsson, I.-S. Magnus and K. Carlström, *J. Biol. Chem.*, **251**, 2831 (1976).
10. D. A. Broadbent and N. J. Cortwright, *Microbios*, **9**, 119 (1974).
11. I. C. Gunsalus, J. R. Meeks, J. D. Lipscomb, P. Debrunner and E. Münck, *Molecular Mechanisms of Oxygen Activation* (ed. O. Hayaishi), p. 559, Academic Press (1974).
12. R. L. Tsai I. C. Gunsalus and K. Dus, *Biochem. Biophys. Res. Commun.*, **45**, 1300 (1971).
13. D. W. Cushman, R. L. Tsai and I. C. Gunsalus, *ibid.*, **26**, 577 (1967).
14. J. Hedegaard and I. C. Gunsalus, *J. Biol. Chem.*, **240**, 4038 (1965).
15. C.-A. Yu and I. C. Gunsalus, *Biochem. Biophys. Res. Commun.*, **40**, 1431 (1970)
16. M. Katagiri, B. N. Ganguli and I. C. Gunsalus, *Fed. Proc.*, **27**, 525 (1968).
17. K. Dus, M. Katagiri, C.-A. Yu, B. L. Erbes and I. C. Gunsalus, *Biochem. Biophys. Res. Commun.*, **40**, 1423 (1970).
18. M. Tanaka, S. Zeitlin, K. T. Yasunobu and I. C. Gunsalus, *Iron and Copper Proteins* (ed. K. T. Yasunobu, H. F. Mower and O. Hayashi), p. 263, Pienuk Publishing (1976).
19. R. L. Tsai, C-A. Yu, I. C. Gunsalus, J. Peisach, W. Blumberg, W. H. Orme-Johnson and H. Beinert, *Proc. Natl. Acad. Sci. U.S.A.*, **66**, 1157 (1970).
20. I. C. Gunsalus, R. L. Tsai, C. A. Tyson M. C. Hsu and C-A. Yu, 4th Int. Conf. Magnetic Resonance in Biological Systems, Oxford, 26, **4**, 75 (1970).
21. Y. Ishimura, V. Ullrich and J. A. Peterson *Biochem. Biophys. Res. Commun.*, **42**, 140 (1971).
22. M. Tanaka, M. Haniu, K. T. Yasunobu, K. Dus and I. C. Gunsalus, *J. Biol. Chem.*, **249**, 3689 (1974).
23. S. Takemori, M. Nakamura, K. Suzuki, M. Katagiri and T. Nakamura, *Biochim. Biophys. Acta*, **284**, 382 (1972).
24. I. C. Gunsalus, C. A. Tyson and J. D. Lipscomb, *Oxidases and Related Redox Systems* (ed. T. E. King, H. S. Mason and M. Morrison), p. 583, University Park Press,

(1973).
25. J. A. Peterson, M. Kusunose, E. Kusunose and M. J. Coon, *J. Biol. Chem.*, **242**, 4334 (1967).
26. J. A. Peterson and M. J. Coon, *ibid.*, **243**, 329 (1968).
27. R. T. Ruettinger S. T. Olson, R. F. Boyer and M. J. Coon, *Biochem. Biophys. Res. Commun.*, **57**, 1011 (1974).
28. Y. H. Anthony, K. W. Lu and M. J. Coon, *J. Biol. Chem.*, **244**, 3714 (1969).
29. C. A. Appleby, *Biochim. Biophys. Acta*, **147**, 399 (1967).
30. D. R. Morris and L. P. Hager, *J. Biol. Chem.*, **241**, 1763 (1966).

4.4 Yeast Microsomal Systems

4.4.1 General Considerations on Yeast Microsomal Electron Transport Systems

The presence of non-mitochondrial hemoproteins in anaerobically grown yeast was already known in the 1930's due to the pioneering work of Euler et al.[1] and of Fink.[2] About 30 years later, non-mitochondrial hemoproteins were reinvestigated by Lindenmayer and Smith[3] in a cell-free system from anaerobically grown yeast. They demonstrated the presence of cytochrome P-450-like CO-binding pigment in the yeast. In 1969, Ishidate et al.[4] reported that the CO-binding pigment and the b-type cytochrome found by Euler and Fink[1,2] and later named cytochrome b_1,[5] are bound to a particulate fraction of the yeast cells, and they also pointed out the apparent similarity of these pigments to the hemoprotein constituents of hepatic microsomes. Since the beginning of the 1970's, these pigments in the particulate fraction of anaerobically grown yeast have been investigated precisely, as postulated components of the yeast microsomal electron transport system, by Yoshida and his co-workers.[6-14] The b-type hemoprotein (hitherto called cytochrome b_1[1-5]) was purified and identified as cytochrome b_5 of yeast.[7,8] The CO-binding pigment was also purified and shown to be cytochrome P-450.[9,10] These findings were soon followed by the isolation of two flavoproteins, designated as NADH-cytochrome b_5 reductase[11] and NADPH-cytochrome c reductase,[12] from the particulate fraction, and the latter was identified as NADPH-cytochrome P-450 reductase.[13] The similarity of these reductases to the corresponding enzymes obtained from hepatic microsomes was also demonstrated.[11-13] Based on the above observations, it has been suggested that the redox components of the particulate fraction of anaerobically grown yeast constitute an electron transport system which resembles the hepatic microsomal system (cf. section 4.1). Actually, it has been demonstrated that cytochrome b_5 and cytochrome P-450 can be readily reduced by both NADH and NADPH when they are bound to yeast microsomes.[4,6]

It has recently been shown by Tamura et al.[14] that the cytochrome b_5-containing system of yeast microsomes functions in the oxidative desaturation reaction of palmitoyl CoA, and they have also demonstrated the presence of a "cyanide-sensitive factor" which serves as a desaturase in the yeast microsomes. Furthermore, it has been suggested that the cytochrome P-450-containing system of yeast contributes to the monooxygenation of certain lipophilic compounds, as demonstrated below. Hence, it may be concluded that the yeast microsomal electron transport system is practically identical to that of hepatocytes not only in its construction but also in its functions.

4.4.2 Purification and Properties of Yeast Cytochrome P-450

A. Purification

Cytochrome P-450 has been purified from two different yeasts, wild-type strain of *Saccharomyces cerevisiae* and strain LM7 of *Candida tropicalis*, by Yoshida and Kumaoka[9,10] and Duppel et al.,[15] respectively. The *S. cerevisiae* cytochrome P-450 was solubilized by sodium cholate from Nagarse-digested microsomes* of anaerobically grown cells, and was purified by ammonium sulfate fractionation followed by chromatography on a column of DEAE-cellulose in the presence of Triton X-100.[10] Purified preparations thus obtained contain 3–4 nmole cytochrome P-450 per mg of protein, and are free from other colored substances.[10] Recently, another purification procedure for the cytochrome of *S. cerevisiae* has been developed (Yoshida, Aoyama, Kumaoka and Kubota, *unpublished results*). This procedure consists of sodium cholate extraction, ammonium sulfate fractionation, chromatography on a column of AH-Sepharose 4B, hydroxylapatite column chromatography, and column chromatography with CM-Sephadex C-50. Since this procedure includes no proteolytic process, it improves greatly on the preceding one.[10] The specific content of preparations obtained by this method is as high as 12–14 nmole of the cytochrome per mg of protein, although some impurities are present in the preparations. A typical result of the purification of yeast cytochrome P-450 by this method and the absorption spectra of the purified preparation are shown in Table 4.7 and Fig. 4.10, respectively.

The cytochrome preparation from *C. tropicalis*[15] has a lesser purity (0.4 nmole cytochrome/mg protein) than that from *S. cerevisiae*, and its

---

* By digestion of the microsomes with Nagarse (a bacterial proteinase), redox components of yeast microsomes other than cytochrome P-450 can be removed from the membranes.

TABLE 4.7. Summary of the purification of cytochrome P-450 from
S. cerevisiae microsomes (Yoshida et al., *unpublished results*)

Step	Protein (mg)	Cytochrome P-450			
		(nmole)	(Yield)	(nmole/mg protein)	(Purification)
Microsomes	1480	98.9	100	0.067	1
Ammonium sulfate fraction	224	55.3	56	0.247	3.69
AH-Sepharose 4B eluate	9.07	54.2	55	5.98	89.3
Hydroxylapatite eluate	3.50	33.4	34	9.54	142
CM-Sephadex C-50 eluate	1.27	16.8	17	13.2	197

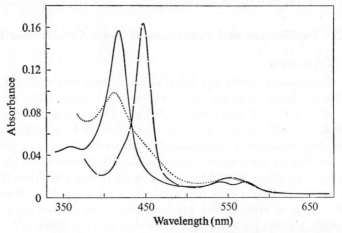

Fig. 4.10. Absorption spectra of cytochrome P-450 from *S. cerevisiae* (Yoshida et al., *unpublished data*). A purified preparation of the cytochrome (12.5 nmole/mg protein) was dissolved in 50 mM potassium phosphate buffer, pH 7.0, containing 20% glycerol. The concentration of the cytochrome was 1.4 μM and the light path length was 1.0 cm. ———, Oxidized form; ······, reduced form with $Na_2S_2O_4$; — —, reduced CO compound.

detailed characterization as a hemoprotein has not yet been made.

B. Properties

Fig. 4.10 shows the absorption spectra of cytochrome P-450 from *S. cerevisiae*. The Soret peak of the oxidized form is situated at 418 nm and no absorption band is observed at around 650 nm, indicating that this form of the cytochrome is in a low spin state (*cf.* section 3.2). However, when the spin state of the cytochrome in the native microsomes was judged according to the method of Jefcoate et al.,[16] it was found that the majority of the cytochrome was in a high spin state. Furthermore, it has been demonstrated[9,10] that the purified low spin preparation of the cytochrome can be converted partially to a high spin state by increasing the ionic strength of the medium, and the reverse change is caused either by an elevation of the concentration of Triton X-100 in the medium or by the addition of alcohols to it. Further, it was found that the high spin cytochrome P-450 in the native microsomes was converted to a low spin state by the addition of Triton X-100 or alcohols. Although no exact interpretation of this is yet available, it is suggested that there may be a tendency for the transition from the high to low spin state of the cytochrome to be caused by the addition of certain compounds such as Triton X-100 and alcohols. It is possible therefore, that the majority of the yeast cytochrome P-450 is in a high spin state when it remains in the native microsomes, but it is converted to an apparent low spin state by the Triton X-100 used in the purification process.

The Soret band of the reduced CO-compound of the cytochrome is situated at 448 nm, not at 450 nm, suggesting a similarity between this cytochrome and the cytochrome P-450 induced by a polycyclic hydrocarbon in hepatic microsomes (called P-448 or P_1-450). It has been reported by Jefcoate et al.[16,17] that high spin cytochrome P-450 is predominant in hepatic microsomes from 3-methylcholanthrene-induced animals, and this high spin cytochrome is converted to an apparent low spin state on solubilization with the non-ionic detergent, Lubrol WX. Furthermore, Fujita et al.[18] have reported that oxidized cytochrome P_1-450 is in an apparent low spin state when it is solubilized with the non-ionic detergent, Triton N-101.

On this basis, it is concluded that the spectral properties of yeast

TABLE 4.8. Spectral parameters† of cytochrome P-450 from *S. cerevisiae* (Yoshida et al., *unpublished results*)

State	λ_{max} (nm)	ε (mM^{-1} cm^{-1})
Oxidized	575	11
	540	11
	418	113
	360	35
Reduced	555	13
	412	71
Reduced CO-compound	555	14
	447	117
Reduced CO-difference	448 – 500	92

† The values of the extinction coefficient were calculated on the assumption that one mole of the cytochrome contains one mole of protoheme. These values differ somewhat from those appearing in a previous report;[10] the present data are more accurate.

cytochrome P-450 closely resemble those of the hepatic cytochrome, especially from polycyclic hydrocarbon-induced animals. The spectral parameters of the purified cytochrome P-450 obtained from *S. cerevisiae* are summarized in Table 4.8.

The prosthetic group of yeast cytochrome P-450 was determined as protoporphyrin IX,[10] as in cases of cytochrome P-450's of different origin. An apparent molecular weight for yeast cytochrome P-450 of about 50,000 daltons was obtained preliminarily by SDS-polyacrylamide gel electrophoresis of the purified preparation.

Yeast cytochrome P-450 binds with ethyl isocyanide in both the oxidized and reduced states.[10] The ethyl isocyanide complex of the reduced cytochrome shows two Soret peaks at 430 and 455 nm, respectively, and the relative intensity of these peaks is dependent on the pH of the medium (Fig. 4.11) as in the case of the hepatic cytochrome (*cf.* section 3.1). This finding indicates that the yeast cytochrome also exists in two interconvertible forms (called the 430 and 455 forms) in the reduced state.

The spectral change induced by the addition of low-molecular orga-

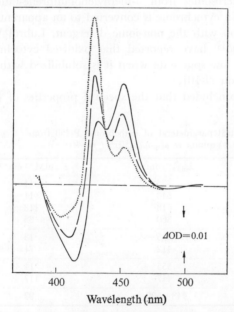

Fig. 4.11. Ethyl isocyanide-difference spectra of cytochrome P-450 from *S. cerevisiae*.[10] A purified preparation of the cytochrome (2.9 nmole/mg protein) was used at a concentration of 0.51 μM. ——, pH 8.0; — —, pH 7.0; ······, pH 6.0.

nic compounds—called substrate-induced spectral change—has also been observed with yeast cytochrome P-450.[10] When the cytochrome is in the microsomes, Type II and modified Type II (or reverse Type I) difference spectra are induced, respectively, by amines and hydroxyl group-containing compounds. Although Type II spectral change is consistently observed, modified Type II spectral change is not observed in the purified preparation unless the ionic strength of the medium is increased.[10] This fact suggests a relationship between the high spin cytochrome and the modified Type II spectral change, as in the case of hepatic microsomes.[19] It should be noted, however, that no Type I spectral change has ever been induced by any of the compounds so far used.[10] A number of barbiturates, the typical Type I-inducing compounds of hepatic cytochrome P-450, have been found to induce a modified Type II spectrum,[10] but the other Type I-inducing compounds of the hepatic cytochrome such as alkanes and higher fatty acids caused no spectral change in the yeast cytochrome.[10]

As in the case of hepatic cytochrome P-450, yeast cytochrome can be denatured to P-420 (Fig. 4.12).[10] Suitable denaturing agents include several organic solvents, high concentrations of detergents, high concentrations of chaotropic anions, and mercurials such as p-chloro-

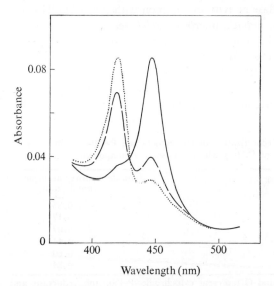

Fig. 4.12. Conversion of cytochrome P-450 to P-420 with p-chloromercuribenzoate. A purified preparation of the cytochrome (2.9 nmole/mg protein) was denatured with 0.1 mM p-chloromercuribenzoate under aerobic conditions. ——, Immediately after the addition of PCMB; — —, 30 min later, ······, 60 min later.

mercuribenzoate and $HgCl_2$. Although the above agents are common with those used for denaturing hepatic cytochrome P-450, the yeast cytochrome is considerably more resistant than the hepatic cytochrome to treatment with non-ionic detergents.

4.4.3 Cytochrome P-450-containing Electron Transport Systems of Yeasts

It has been reported by Duppel et al.[15] that the cytochrome P-450 of *C. tropicalis* constitutes an electron transport system together with its reductase that requires NADPH as the electron donor. The reductase shows NADPH-cytochrome *c* reducing activity,[15] so resembling the NADPH-cytochrome P-450 reductase of hepatic microsomes. On reconstitution of the electron transport system from its isolated components, a lipid factor was required other than the cytochrome and the reductase (Table 4.9),[15] but no iron-sulfur protein such as adrenodoxin or putidaredoxin was required.[15] Recently, Aoyama et al.[13] isolated the *S. cerevisiae* NADPH-cytochrome *c* reductase which retains an amphipatic nature. This reductase contains one mole each of FAD and FMN as its prosthetic groups, and shows menadion (vitamin K_3)-dependent NADPH oxidase activity in common with the NADPH-cytochrome P-450 reductase of hepatic microsomes (*cf.* section 4.1). Furthermore, this reductase readily reduces isolated cytochrome P-450 of yeast in the presence of a trace amount of detergent or phospholipid isolated from the yeast microsomes.[13] It can therefore be concluded that the constitution of the cytochrome P-450-containing electron transport chain of *S. cerevisiae* microsomes is practically identical with that of hepatic microsomes.

TABLE 4.9. Reconstitution of the cytochrome P-450-containing system of *C. tropicalis* from isolated components

Fractions present[1]	Laurate hydroxylation activity[2] (nmole/min)
A	0.35
B	0.17
C	0.07
A + B	0.44
A + C	0.60
A + B + C	0.84

[1] Fractions A, B and C represent cytochrome P-450, the reductase and lipid factor, respectively.
[2] Contribution of the cytochrome P-450-containing system to this activity was established.
(Source: ref. 15. Reproduced by kind permission of the Federation of European Biochemical Societies, W. Germany.)

Cytochrome P-450-containing electron transport systems can be classified into two categories: one includes an iron-sulfur protein as the intermediate electron carrier, and the other does not. The former type occurs in adrenocortical mitochondria (*cf.* section 4.2) and in *Pseudomonas putida* grown on camphor (*cf.* section 4.3) and is characterized by its incomplete integration into the membranes. The latter type occurs typically in hepatic microsomes (*cf.* section 4.1) and is integrated firmly into the membranes. It is also recognized that the reductase of hepatic microsomes contains one mole each of FAD and FMN and exhibits cytochrome *c* reductase and menadion-dependent NADPH oxidase activities. On the other hand, the flavoproteins of adrenocortical mitochondria and of *Ps. putida* systems contain a single species of flavin prosthetic group and show neither cytochrome *c* reductase nor menadion-dependent NADPH oxidase activity. Based on the above-mentioned properties of yeast microsomal systems and on the above general criteria, the cytochrome P-450-containing systems of yeasts should be included under the same category as the hepatic microsomal systems.

4.4.4 Function of Cytochrome P-450-containing Systems of Yeasts

As described in Chapter 2, most of the cytochrome P-450-containing systems of various organisms are known to serve as biological catalysts for the monooxygenation of a wide variety of substances. It has been reported by Coon and his co-workers[15,20] that the cytochrome P-450-containing system of *C. tropicalis* is also functional in the monooxygenation of hydrocarbons, fatty acids and drugs. Such catalytic features of the *C. tropicalis* system, especially its low substrate specificity for lipophilic compounds, resemble to a surprisingly extent the hepatic cytochrome P-450-containing system. It should be noted, however, in the case of the *C. tropicalis* system that it has been observed only in cells grown on hydrocarbons under aerobic conditions.[20] Under these conditions, the system must play an important role in the oxidative metabolism of the main carbon source, hydrocarbon. Therefore, it should be considered that the system in *C. tropicalis* contributes directly to the growth of the cells, just as in the case of the similar system in *Ps. putida*. It has been suggested by Alexander *et al.*[21] that the conversion of lanosterol to zymosterol (and also to ergosterol) in cells of *S. cerevisiae* may be mediated, at least a part, by a system containing cytochrome P-450. Although this possibility was inferred simply from the observed inhibitory effect of CO on lanosterol metabolism, the contribution of the cytochrome to sterol biogenesis in the yeast appears likely from an analogy with the cholesterol synthesis occurring in hepatocytes (*cf.* Chapter 2). In addition, it is known that the content of cytochrome P-450 in cells of *S. cerevisiae* varies according to the growth conditions of the cells, as described in Chapter 6. Therefore, a further precise assessment is required to decide the exact function of cytochrome P-450 based on physiological changes in the cytochrome.

References

Section 4.4
1. H. von Euler, H. Fink and H. Hellstrem, *Hoppe-Seyler's Z. Physiol. Chem.*, **169**, 10 (1927).
2. H. Fink, *ibid.*, **210**, 197 (1932).
3. A. Lindenmayer and L. Smith, *Biochim. Biophys. Acta*, **93**, 445 (1964).
4. K. Ishidate, K. Kwaguchi, K. Tagawa and B. Hagihara, *J. Biochem. (Tokyo)*, **65**, 375 (1969).
5. B. Ephrussi and P. P. Slonimsky, *Biochim. Biophys. Acta*, **6**, 625 (1950).
6. Y. Yoshida, H. Kumaoka and R. Sato, *J. Biochem. (Tokyo)*, **75**, 1201 (1974).
7. Y. Yoshida and H. Kumaoka, *Biochim. Biophys. Acta*, **189**, 461 (1969).
8. Y. Yoshida, H. Kumaoka and R. Sato, *J. Biochem. (Tokyo)*, **75**, 1211 (1974).
9. Y. Yoshida and H. Kumaoka, *ibid.*, **71**, 915 (1972).
10. Y. Yoshida and H. Kumaoka, *ibid.* **78**, 785 (1975).
11. S. Kubota, Y. Yoshida and H. Kumaoka, *ibid.*, **81**, 187 (1977).
12. S. Kubota, Y. Yoshida, H. Kumaoka and A. Furumichi, *ibid.*, **81**, 197 (1977).
13. Y. Aoyama, Y. Yoshida, S. Kubota and H. Kumaoka, *Seikagaku* (Japanese), **47**, 506 (1975).
14. Y. Tamura, Y. Yoshida, R. Sato and H. Kumaoka, *Arch. Biochem. Biophys.*, **175**, 284 (1976).
15. W. Duppel, J.-M. Lebeault and M. J. Coon, *Eur. J. Biochem.*, **36**, 583 (1973).
16. C. R. E. Jefcoate, J. L. Gaylor and R. L. Calabrese, *Biochemistry*, **8**, 3455 (1969).
17. C. R. E. Jefcoate and J. L. Gaylor, *ibid.*, **8**, 3464 (1969).
18. T. Fujita, D. W. Shoeman and G. J. Mannering, *J. Biol. Chem.*, **248**, 2192 (1973).
19. Y. Yoshida and H. Kumaoka, *J. Biochem. (Tokyo)*, **78**, 455 (1975).
20. J. M. Lebeault, E. T. Lobe and M. J. Coon, *Biochem. Biophys. Res. Commun.*, **42**, 413 (1971).
21. K. T. W. Alexander, K. A. Mitropoulous and G. F. Gibbons, *ibid.*, **60**, 460 (1974).

4.5 Induction and Disappearance of Cytochrome P-450 in Yeast Cells

The occurrence of cytochrome P-450 in yeast cells was first reported by Lindenmayer *et al.*[1] Previously, in animal tissues, the cytochrome had been investigated by a number of workers and its biochemical properties as well as physiological role in the detoxication and synthesis of sterol compounds were established. Unfortunately, however, the corresponding situation in yeast cells has been studied in less detail. Cytochrome P-450 is widely distributed and plays multiple roles in various kinds of yeast cells. Furthermore, in facultatively aerobic yeast cells, it shares the microsomal membrane with cytochrome b_1 (yeast cytochrome b_5) and reductases transferring electrons from NADH or NADPH, as in mammalian hepatocytes.[2-5] Therefore, an understanding of the induction mechanism and physiological roles of the cytochrome may help to elucidate not only the metabolism of compounds catalyzed by this cytochrome in yeast cells, but details of the membrane-bound electron transport system in general. In this chapter, changes in the cytochrome content among yeast cells grown under various conditions are described, and their physiological significance is discussed in terms of the role of the cytochrome in sterol metabolism.

4.5.1 Induction of Cytochrome P-450 in Alkane-utilizing Yeast

Some species of yeasts are known to utilize hydrocarbons for growth. It has been reported by two groups[6,7] that cytochrome P-450 is an essential component of the alkane-oxidizing enzyme system of *Candida tropicalis*, a strictly aerobic yeast. Lebeault et al.[7] indicated that the enzyme system is also capable of performing the oxidation of fatty acids and demethylation of various drugs in the presence of NADPH and oxygen. In this case, neither enzyme activity nor cytochrome P-450 is recovered in particulate fractions, but recovery is from the soluble fraction when the cells are disintegrated by a French Press. On the other hand, Gallo et al.[6] claimed the localization of both hydroxylase activity and the cytochrome in the microsomal fraction. One consistent feature in both cases, however, is that the extracts contain a relatively large amount of the cytochrome but it is inducible and not detected unless the cells are grown on a hydrocarbon.

By employing various hydrocarbon-utilizing strains of *Candida* yeasts, the morphological changes brought about by the introduction of hydrocarbons have been investigated by Osumi et al.[8,9] They have observed that in hydrocarbon-adapted cells, irrespective of the strain, several microbodies are formed with a concomitant rise in the catalase activity level, whereas the organelles are scarcely found and the level of the enzyme is very low in glucose-grown cells. Unfortunately, no observations were carried out by this group on the cytochrome P-450 content, although it is likely that the induced alkane-oxidizing system including cytochrome P-450 is confined to the newly formed microbodies in the hydrocarbon-utilizing yeast cells. The organelles appear to be broken up on disruption of the cells by a French Press, and subsequently the cytochrome present in them may leak out and be recovered in the soluble fraction, or in the microsomal fraction if tightly bound to the membrane. If this is in fact so, the intracellular localization as well as induction of the cytochrome will present new interesting problems in yeast cytology.

4.5.2 Changes in the Cytochrome P-450 Content of Facultatively Aerobic Yeast Cells Grown under Various Conditions

No alkane-oxidizing system has yet been identified in facultatively aerobic yeast cells such as *Saccharomyces*, in which the cytochrome P-450 is bound tightly to the membranes and never solubilized by mechanical disintegration of the cells. The electron transport system involving cytochrome P-450 in *Saccharomyces cerevisiae* is similar in many respects to that of mammalian liver microsomes. However, although the latter system is known to play a significant role in drug hydroxylation and to be induced by the administration of certain kinds of drugs, such as phenobarbital and methylcholanthrene, there is no report describing the induction of cytochrome P-450 in yeast species, except for a recent brief communication.[10] This paper reports that a small amount of cytochrome P-450 is inductively produced by the addition of phenobarbital to a medium in which the glucose content is as low as 0.5% and no cytochrome P-450 is synthesized. The drug, however, exhibited no effect on the production of the cytochrome in media containing high concentrations of glucose. On the other hand, the cellular content of the cytochrome in cells of *Saccharomyces cerevisiae* has been shown to vary over a wide range, depending on the conditions of growth.[2] As shown in Table 4.10, the cytochrome is found in appreciable amounts in anaerobically grown cells, but only slightly in aerobic cells. The data also indicate that the cytochrome content is not significantly affected by mitochondrial gene mutations. In cells of the petite mutant yeast, which lacks mitochondrial respiration, the cytochrome content is approximately

TABLE 4.10. Content of cytochrome P-450 in various kinds of yeast cells[2]

Strain	Growing conditions	Cyt. P-450 content (nmole/10^{10} cells)
Wild-type	Anaerobic	0.93
	Semi-anaerobic	7.73
	Aerobic	trace
Petite mutant	Semi-anaerobic	8.90
	Aerobic	0.10

as high as that in wild-type yeast cells grown under identical conditions. The mutant yeast cells have been used very conveniently for spectrophotometric studies of the cytochrome, since they lack cytochrome oxidase which exhibits a negative absorption at around 450 nm in the CO-difference spectrum.

Not only oxygen, but also the carbon source and its concentration exert an influence on the cytochrome content of yeast cells. Thus, even in aerobic cells, a considerable amount of cytochrome P-450 is found when they are grown in the presence of high concentrations of glucose and harvested at the late exponential phase of growth, while there is only a trace amount of the cytochrome in cells harvested at the late stationary phase, where the glucose in the culture medium is exhausted, or when they are grown on a non-fermentable carbon source such as lactate.

Based on these observations, it can be said that cytochrome P-450 is present at high levels in cells grown under anaerobic conditions or in the presence of high concentrations of glucose. This may indicate an inverse relationship between the cellular content of the cytochrome and mitochondrial repiratory activity, since both environmental factors are known to be very repressive towards the formation of the mitochondrial respiratory system. Actually, it has been reported that the cytochrome content decreases rapidly on aerobic adaptation of the anaerobic yeast cells, and that this decrease is attenuated in the presence of chloramphenicol, a potent inhibitor of mitochondrial protein synthesis, or of high concentrations of glucose.[11] However, it appears that the inverse relationship between cytochrome P-450 content and mitochondrial respiration may be accidental rather than inherent, since a somewhat slower but similar decrease in cytochrome content was observed during aeration of anaerobic petite mutant cells in which the mitochondrial respiratory system is never formed. Furthermore, when semi-anaerobic cultivation was carried out with galactose as a carbon source (this, unlike glucose, does not repress the formation of the mitochondrial respiratory system[12]), the level of the cytochrome, in spite of a strong mitochondrial respiratory activity, was found to be approximately as high as that in glucose-grown cells (K. Tagawa, *unpublished results*). Thus, it appears that the rapid decrease in cytochrome P-450 content during aeration is not directly related to the formation of the mitochondrial respiratory system, but may be caused by some morphological change in the intracellular membrane system binding the cytochrome.

Since the cytochrome is found in large quantities in semi-anaerobic cells, it may be expected that the cytochrome is induced when aerobic cells are incubated in the presence of high concentrations of glucose under

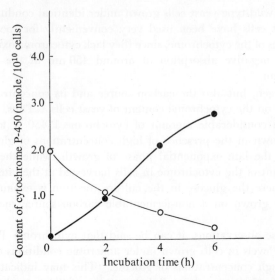

Fig. 4.13. Induction and disappearance of cytochrome P-450 in petite mutant yeast cells. The mutant cells were grown aerobically and harvested at the late exponential (18 h) and the late stationary (32 h) phases. The exponential cells were incubated semi-anaerobically in a non-proliferating medium consisting of 4% glucose and 0.025 M potassium phosphate, pH 6.5 (○), while the stationary cells were incubated semi-anaerobically in a growing medium consisting of 1% yeast extract, 1% polypeptone and 4% glucose (●).

semi-anaerobic conditions. Recent studies in our laboratory have revealed, however, that induction of the cytochrome does not occur under any conditions in a non-proliferating medium devoid of a nitrogen source, in which mitochondrial cytochromes are nevertheless adaptively synthesized under aerobic conditions; induction of the cytochrome is performed only in growing cells (Fig. 4.13). Furthermore, the cytochrome content was found to decrease gradually even under semi-anaerobic conditions when the cells containing the cytochrome were incubated in the non-proliferating medium. Recently, Wiseman et al.[13] have reported that in an aerobic culture of brewer's yeast, cytochrome P-450 is produced during growth at higher concentrations of glucose than 1% but it is rapidly degraded at the stationary phase, and that a large (more than double) amount of cytochrome P-450 is found at the stationary phase when 3% Tween 80 is added into the medium containing 20% glucose, probably due to a lowered rate of degradation rather than an inductive effect. Although the inductive mechanism of the cytochrome in yeast cells has not yet been clarified, these authors suggest that the production of the cytochrome is controlled by the cellular level of cyclic AMP, which is lowered in the presence of high concentrations of glucose.

Summarizing the above observations, it can be concluded that cytochrome P-450 is synthesized during growth in a medium containing a fermentable sugar and disappears when growth ceases, and that the rate of synthesis is rather greater in the presence of molecular oxygen than in its absence, while the rate of disappearance is also accelerated under aerobic conditions.

4.5.3 Physiological Significance of Changes in Cytochrome P-450 Content

The possible involvement of cytochrome P-450 in ergosterol synthesis has been shown by the fact that the oxidative demethylation of lanosterol by a cell-free preparation obtained from semi-anaerobic yeast cells is partially inhibited by CO.[14] Before this, inhibition by CO had also been observed in the cholesterol synthesis of rat liver microsomes.[15] The above possibility is also supported by the observation that mutant yeast cells defective in porphyrin biosynthesis accumulate lanosterol.[16] However, these observations are not consistent with the high level of the cytochrome in anaerobic cells or its rapid fall upon aeration, because ergosterol synthesis requires molecular oxygen and aerobic cells attain an extremely high level of the lipid.[17] Furthermore, ergosterol production has been found to continue during the aerobic adaptation of anaerobic yeast cells, even after the cytochrome has become undetectable (K. Ishidate and K. Tagawa, *unpublished results*). Thus, the change in cytochrome P-450 content has not been reasonably explained, and the following possibility is therefore presented. It is feasible that cytochrome P-450 does not actually disappear in the aerated cells in a non-proliferating medium but becomes undetectable from the usual CO-difference spectrum due to changes in its oxidation-reduction state and binding affinity for CO during prolonged aeration. This idea is apparently supported by the following observations: (1) the oxidative demethylation activity of lanosterol in cell-free preparations is inhibited only partially by CO,[14] (2) the level of ergosterol in semi-anaerobic cells is unaffected by CO,[3] and (3) in the oxygen uptake by membrane preparations involving cytochrome P-450 obtained from semi-anaerobic petite mutant yeast cells, the value of the inhibitory constant (K_i) for CO obtained kinetically is about 10 times as large as the dissociation constant (K_s) of the CO-complex of the cytochrome obtained spectrophotometrically.[3] The above possibility, however, is merely speculative at present, and the mechanism for the apparent decrease in cytochrome P-450 content must be examined fully in the future.

REFERENCES

Section 4.5
1. A. Lindenmayer and L. Smith, *Biochim. Biophys. Acta*, **93** 445 (1964).
2. K. Ishidate, K. Kawaguchi, K. Tagawa and B. Hagihara, *J. Biochem. (Tokyo)*, **65**, 375 (1969).
3. K. Kawaguchi, K. Ishidate and K. Tagawa, *ibid.*, **74**, 817 (1973).
4. Y. Yoshida and H. Kumaoka, *ibid.*, **71**, 915 (1972).
5. Y. Yoshida, H. Kumaoka and R. Sato, *ibid.*, **75**, 1201 (1974).
6. M. Gallo, J. C. Bertrand and E. Azoulay, *FEBS Lett.*, **19**, 45 (1971).
7. J. M. Lebeault, E. T. Lode and M. J. Coon, *Biochem. Biophys. Res. Commun.*, **42**, 413 (1971).
8. M. Osumi, N. Miwa, Y. Teranishi, A. Tanaka and S. Fukui, *Arch. Microbiol.*, **99**, 181 (1974).
9. M. Osumi, F. Fukuzumi, Y. Teranishi, A. Tanaka and S. Fukui, *ibid.*, **103**, 1 (1975).
10. A. Wiseman and T.-K. Lim, *Biochem. Soc. Trans.*, **3**, 974 (1975).
11. K. Ishidate, K. Kawaguchi and K. Tagawa, *J. Biochem. (Tokyo)*, **65**, 385 (1969).
12. I. Nagata, Y. Yoshida, E. Furuya, T. Kanaseki and K. Tagawa, *ibid.*, **78**, 1535 (1975).
13. A. Wiseman, C. McCloud and T.-K. Lim, *Biochem. Soc. Trans.*, **4**, 685 (1976).
14. K. T. W. Alexander, K. A. Mitropoulos and G. F. Gibbons, *Biochem. Biophys. Res. Commun.*, **60**, 460 (1974).
15. G. F. Gibbons and K. A. Mitropoulos, *Eur. J. Biochem.*, **40**, 267 (1973).
16. M. Bard, R. A. Woods and J. M. Haslam, *Biochem. Biophs. Res. Commun.*, **74**, 324 (1974).
17. D. Jollow, G. M. Kellerman and A. W. Linnane, *J. Cell Biol.*, **37**, 221 (1968).

CHAPTER 5

Mechanisms of Cytochrome P-450-catalyzed Reactions[*1]

5.1. Introduction
5.2. Substrate Interaction with Cytochrome P-450
5.3. Oxygen Interaction with Cytochrome P-450
5.4. The Role of Iron-Sulfur Proteins
5.5. Reaction Mechanism of P-450$_{CAM}$ and Reactive Species of Oxygen
5.6. Problems in Cytochrome P-450 Systems of Mammalian Tissues

5.1 INTRODUCTION

As has been discussed in previous chapters,[1-3] cytochrome P-450 represents a family of protohemoproteins that is widely distributed in nature. It has been shown to be responsible for a variety of monooxygenase reactions, the substrates of which include various physiologically important substances such as steroids, hydrocarbons, lipids and xenobiotics. It is a unique monooxygenase containing heme as the oxygen-activating site, and no other hemoprotein is known that catalyzes the monooxygenase reactions.[*2] Cytochrome P-450 is also unique in that it shows peculiar spectroscopic properties, as exemplified by the optical spectrum of the carbon monoxide-ligated ferrous form and the EPR spectrum of the ferric form. Thus, a considerable number of reports has appeared concerning the function, structure and mechanism of action of this class of cytochromes.[5-11] However, precise analyses of the mechanism of action, particularly in the case of mammalian cytochrome P-450's have not been successful until rather recently, due mainly to the particulate nature of the systems. Solubilization and purifica-

---

[*1] Yuzuru ISHIMURA, Department of Biochemistry, Keio University School of Medicine, Shinjuku-ku, Tokyo 160, Japan
[*2] Peroxidase catalyzes the hydroxylation of certain aromatic compounds such as salicylic acid and nitrobenzene in the presence of dihydroxyfumaric acid and oxygen.[4]

tion of the cytochromes, either from microsomes or mitochondria, was greatly hampered by their delicate instability. During the course of such treatments, the cytochrome P-450 is liable to be converted to cytochrome P-420, which is catalytically inactive and is considered to represent a denatured form of cytochrome P-450.[4,12] In 1968, however, Katagiri et al.[13] found that a D-camphor hydroxylating system of *Pseudomonas putida* contained a soluble cytochrome P-450; this, they subsequently purified to a homogeneous state. They were also able to show that the soluble cytochrome P-450 acted as the terminal monooxygenase for a system in which a flavoprotein (putidaredoxin reductase) and an iron-sulfur protein (putidaredoxin) served as electron-transferring components, thereby permitting extensive analysis of the reaction mechanism.

The bacterial cytochrome designated by Katagiri et al. as $P-450_{CAM}$, was found to share almost all the spectral as well as catalytic properties of membrane-bound cytochrome P-450, as follows. (1) When reduced and ligated with carbon monoxide, it exhibited a Soret absorption maximum near 450 nm, an unusually long wavelength for a carbon monoxide complex of a protohemoprotein.[13] (3) Upon addition of D-camphor, a so-called substrate-induced spectral change could be observed.[13] (2) It required a specific electron transfer system, NADH-putidaredoxin reductase and putidaredoxin, for the reaction in a manner similar to the requirement for NADPH-adrenodoxin reductase and adrenodoxin in the cytochrome P-450 system obtained from adrenal cortex mitochondria.[13] (4) A unique set of EPR signals at around $g=2.4$, 2.2 and 1.9 was observable with the substrate-free ferric form, whereas the substrate-bound form exhibited signals at around $g=8$ and 4 at cryogenic temperatures.[14] In addition, a recent report from Coon's laboratory[15] has indicated a close similarity in protein structure between the cytochrome P-450 of liver microsomes and $P-450_{CAM}$, as judged by both amino acid analysis and the immunochemical cross reactions. Thus, the bacterial system is currently being employed as an appropriate model for the elucidation of the general reaction mechanisms of cytochrome P-450-catalyzed reactions. The stoichiometry of the overall reaction is shown in Eq. (5.1).

$$NADH + H^+ + O_2 + camphor \rightarrow camphor\text{-}OH + NAD^+ + H_2O \quad (5.1)$$

The purpose of this chapter is to review the present status of investigations on the mechanisms of cytochrome P-450-catalyzed reactions, with particular emphasis on $P-450_{CAM}$. The discussion will be centered mainly on the nature and reactivities of the three stable reaction intermediates so far isolated, viz. the ferric $P-450_{CAM}$-camphor complex, its reduced form, and the oxygen-ferrous P-450$_{cam}$-camphor complex. The data cited are derived mainly from the reports of Gunsalus and his associates, and from those of Peterson et al., as well as our own works. An attempt is also made to give some insight into the mechanisms of oxygen activation in cytochrome P-450-catalyzed monooxygenation reactions.

5.2 Substrate Interaction with Cytochrome P-450

Since the initial report of Narasimhulu et al.[16] that significant changes in the optical spectra of adrenal cortex microsomes suspensions were observable upon the addition of 17α-hydroxyprogesterone, similar spectral findings have been made with various cytochrome P-450-containing systems, such as liver microsomes and adrenal cortex mitochondria, with the respective substrates.[17-19] Such spectral changes, termed substrate-induced spectral changes, have been assumed to reflect an interaction of the substrates with cytochrome P-450. In the case of liver microsomes, it has been possible to show a parallel relationship between the magnitude of the changes and the concentration of cytochrome P-450 in the microsomal membranes.[20] The concentration of added substrate required for half maximal changes in the spectra was often found to coincide with the Michaelis constant for a monooxygenase reaction with the substrate.[20] Changes in EPR spectra were also observable.[21-23] Meanwhile, it was shown that the addition of substrates to the reduced state of a system containing cytochrome P-450 resulted in a lesser degree of spectral change, while treatment of the system with ferricyanide had no appreciable effect on the magnitude of the substrate-induced spectral changes.[24] Thus, Estabrook et al.[24] as well as Sato[4] proposed a reaction mechanism in which the ferric form of cytochrome P-450 primarily interacts with substrates to form an enzyme-substrate complex, and is then reduced to a ferrous state by the action of electron transfer components in the system. The reaction scheme, as initially postulated by Estabrook et al, is shown in Fig. 5.1. The initial binding of substrates to ferric cytochrome P-450 was later supported by the fact that solubilized and partially purified preparations of cytochrome P-450 from liver microsomes show identical substrate-induced spectral changes to the native microsomes.[25] However, the most convincing evidence has become available through the use of soluble cytochrome P-450 of bacterial origin, i.e. P-450$_{CAM}$.

Katagiri, Gunguli and Gunsalus,[13] when they first discovered the

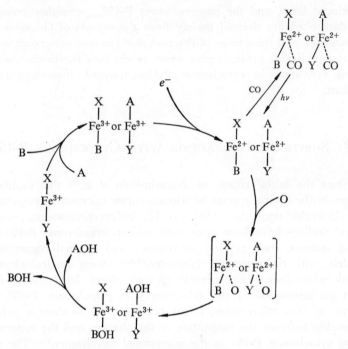

Fig. 5.1. Hypothetical reaction schema for cytochrome P-450-catalyzed reactions as postulated in 1964.[24)] X and Y represent the protein ligands associated with the heme. Substrates causing Type I and Type II spectral changes are denoted A and B, respectively.

participation of cytochrome P-450 in the methylene hydroxylation of D-camphor, noticed significant changes in the optical properties of P-450_{CAM} upon the addition of D-camphor. The optical spectrum of ferric P-450_{CAM} in the camphor-free state, which is characterized by visible bands at 417, 539 and 569 nm, changed into one with absorption maxima at 391, 500 and 640 nm upon the addition of D-camphor. In the difference spectra, the spectral change was marked by new peaks at 388 and 505 nm, as well as troughs at 419 and 570 nm, respectively. Similar changes in the difference spectra had also been described with adrenal cortex mitochondria and deoxycorticosterone as the added substrate.[26)] Subsequently, the EPR experiments of Tsai et al.[14)] gave the following results. (1) The substrate-free form of P-450_{CAM} exhibited g values of 2.46, 2.26 and 1.91, which are characteristic of a heme iron in a low spin ferric state. (2) On addition of D-camphor in saturated amounts, signals due to high spin ferric iron appeared at $g=8$, 4 and 1.8, with a concomitant decrease in the low spin signals. In this study, the high spin signal accounted for about 60% of the total ferric heme, whereas about 40% remained in the low spin state. On the other hand, Peterson,[27] with the aid of Iizuka, was able to show that 75% of the total heme iron was convertible to the high spin ferric state by the binding of D-camphor, as judged from the magnetic susceptibility between 94–253 K. These low temperature EPR and magnetic susceptibility data have recently been confirmed by Mössbauer spectroscopy.[28] However, it remains to be determined whether the low spin fractions remaining after D-camphor addition are due to a temperature-dependent spin equilibrium or are simply indicative of a denatured fraction in the preparation.

Equilibrium and kinetic studies on the binding of D-camphor have been carried out by Peterson and Griffin.[27,29] Based on the spectral shift, the equilibrium association constant for camphor binding was estimated to be 4.7×10^5 M^{-1} in 20 mM potassium phosphate buffer (pH 7.4) containing 0.1 M potassium chloride.[27] They also found that the binding of D-camphor is remarkably enhanced in the presence of cations, with K^+ representing a preferred ion. Thus, the observed K_s for the camphor binding varied in the range of 1–5 μM, depending on the precise experimental conditions such as buffer and specific ion concentrations. The association reaction between P-450$_{CAM}$ and D-camphor was found to be second order with a rate constant of 4.1×10^6 $M^{-1}.sec^{-1}$ at 4°C, while the reverse reaction exhibited first order kinetics with a rate constant of 6.0 sec^{-1}.[29] Although the above measurements were carried out at 4°C to slow down the reaction rate within the limits of the stopped flow technique, the association rate constant obtained was sufficiently large for the binding reaction to be part of the overall monooxygenation reaction. The molecular activity of P-450$_{CAM}$ under optimum conditions at 24°C was reported to be 10–20 sec^{-1}.[29,30]

The binding of D-camphor with P-450$_{CAM}$ and the associated spin state conversion described above are accompanied by several important modifications in enzymatic properties. Noteworthy among them, is the significant shift in redox potential of the heme iron. According to Gunsalus, et al.[30] the E_0' of low spin ferric P-450$_{CAM}$ is -270 mV, while that of the high spin ferric form (enzyme-camphor complex) is -170 mV. Since the E_0' of putidaredoxin is -240 mV, spin state conversion results in a facilitation of heme reduction, thereby enabling successive binding of oxygen to the heme iron. It is a well-known fact that oxygen combines exclusively with the ferrous form of hemoproteins. Changes in enzyme conformation have also been demonstrated by means of various techniques such as CD and MCD, and by examining the stability of the protein against denaturation.[27,30–32]

The competitive relationship between camphor binding and the binding of various ligands such as imidazole and metyrapone to the heme is also noteworthy.[29,30,33] Although the binding site of D-camphor in cytochrome P-450$_{CAM}$ has not yet been established, the competitive nature of camphor binding with respect to ligand binding may perhaps imply that camphor binds with the protein in the proximity of the heme. By examining the binding of various camphor analogs to P-450$_{CAM}$, Gunsalus et al.[34] have indicated that the presence of the carbonyl group of camphor is required for the substrate binding. The substrate interaction of mammalian cytochrome P-450 will be discussed later in this chapter.

5.3 Oxygen Interaction with Cytochrome P-450

Since one of the most important functions of heme in nature is to combine reversibly with oxygen, the binding site for oxygen in cytochrome P-450 has long been assumed to be the heme in the enzyme, although direct experimental evidence for this was not available until 1971.[35-38] The ferric form of cytochrome P-450$_{CAM}$, in the presence of D-camphor, can be reduced either enzymatically by NADH in the presence of putidaredoxin and its reductase, as mentioned above, or chemically by the addition of dithionite. The ferrous cytochrome P-450$_{CAM}$ thus obtained in the presence of D-camphor has been shown to be in a high spin state, as judged by high resolution proton nuclear magnetic spectroscopy.[39] When oxygen was added to the reduced enzyme either by gassing or by the addition of oxygen in solution, the formation of a new spectral species was observed immediately. This had absorption maxima at 418 and 555 nm, with shoulders at 360 and 580 nm (Fig. 5.2). The new spectral species was rather stable in the presence of D-camphor and could be reversibly transformed into CO-cytochrome P-450 by varying the partial pressures of oxygen and carbon monoxide. No EPR signal attributable to the new species was detected between 500 and 4000 gauss at either 10.5 or 77 K, suggesting that the new spectral entity was diamagnetic.[35] These findings, together with the spectral characteristics shown in Table 5.1, indicate that the new species is a complex of oxygen with the reduced heme of the cytochrome P-450. Subsequently, the presence of a similar spectral species during the aerobic steady state of substrate hydroxylation by rat liver microsomes was recognized by Estabrook et al.,[44] and has recently been confirmed by the findings of several laboratories.[45,46] Schleyer et al.[47] reported also that

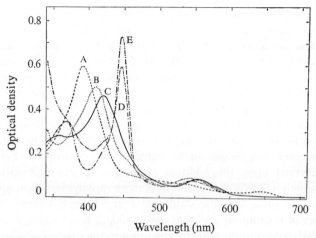

Fig. 5.2. Absorption spectra of various forms of P-450$_{CAM}$ in the presence of D-camphor. A, High spin ferric form; B, ferrous form; C, oxygenated form; D, carbon monoxide form obtained by bubbling CO for 30 sec into C; E, carbon monoxide form obtained after addition of an excess of sodium dithionite to D and bubbling again with CO for 30 sec.
(Source: ref. 35. Reproduced by kind permission of Academic Press, Inc., U.S.A.)

TABLE 5.1. Absorption maxima of oxygenated hemoproteins

	Absorption maxima (nm)			Ref.
	γ	β	α	
Oxyhemoglobin	415	541.5	576	40
Oxymyoglobin	418	544	582	41
Oxyperoxidase	418	546	581	42
Oxygenated intermediate of tryptophan 2,3-dioxygenase	418	545	580	43
New species of cytochrome P-450	418	555	(580)	35

they detected the presence of an oxygenated form of cytochrome P-450 with partially purified preparations from adrenal cortex mitochondria, although detailed information on the results is not available.

The formation of oxy-cytochrome P-450$_{CAM}$, as judged by the stopped flow technique, has been found to obey first-order kinetics with respect to both oxygen and the reduced form of the enzyme, and the second-order rate constant for the reaction at pH 7.4 was shown to be 7.7×10^5 M^{-1}·sec^{-1} at 4°C.[29] The reverse rate constant, i.e. the dissociation rate constant of oxygen from oxy-P-450$_{CAM}$, was estimated to be 1.1 sec^{-1} under similar conditions. No detectable intermediate was observed during either the association or dissociation process. The oxy-P-450$_{CAM}$ thus formed was found to be rather stable at 4°C and neutral pH in the presence of camphor, although it is known to undergo a slow spontaneous transformation, regenerating the ferric form of P-450$_{CAM}$, showing clear isosbestic points. The slow degradation process was first order with respect to the oxy-P-450 concentration, and was accelerated by an increasing hydrogen ion concentration and/or temperature.[29] The rate constants at pH 7.4 were found to be 3.2×10^{-4} and 4.3×10^{-3} sec^{-1} at 4 and 23°C, respectively. On the other hand, the addition of oxygen to ferrous P-450$_{CAM}$ in the absence of D-camphor did not result in the formation of such a stable oxygen adduct. According to Gunsalus, et al.[30] however, the formation of transient oxy-cytochrome P-450$_{CAM}$ can be observed upon the addition of oxygen to ferrous cytochrome P-450, although it decays nearly 10^2 times more rapidly. Hence it, appears reasonable to assume that the stable oxy-cytochrome P-450 obtained in the presence of D-camphor and that oxy-P-450$_{CAM}$ is not a simple adduct of molecular oxygen to the cytochrome but a ternary complex of oxygen, D-camphor and cytochrome P-450. It should also be noted here that no hydroxylated product of D-camphor is formed during the auto-decomposition of the stable oxy-cytochrome P-450, a ternary complex of oxygen, camphor and the enzyme. The mechanism for the dismutation process, especially regarding the fate of the oxygen and one reducing equivalent, still remains to be elucidated. As will be discussed below, the stable oxy-P-450$_{CAM}$ is degraded instantaneously by the addition of reduced putidaredoxin giving the reaction product, 5-*exo*-hydroxycamphor.

5.4 The Role of Iron-Sulfur Proteins

The electron transfer system required for the camphor hydroxylation catalyzed by cytochrome P-450$_{CAM}$ is shown in Fig. 5.3. The system consists of NADH-putidaredoxin reductase (a flavoprotein), putidaredoxin (an iron-sulfur protein), and the cytochrome P-450.[13] A similar electron transfer system composed of a flavoprotein, an iron-sulfur protein (adrenodoxin) and a cytochrome P-450 has been shown to be functional

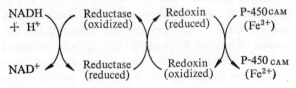

Fig. 5.3. Electron transfer system of P-450$_{CAM}$.

in the steroid hydroxylation system present in adrenal cortex mitochondria.[48)]

The NADH-putidaredoxin reductase is a flavoprotein with a molecular weight of 43,500 containing one mole of FAD per mole of enzyme.[38)] The flavoprotein with $E_0' = -285$ mV undergoes a two electron reduction by NADH and reduces the oxidized form of putidaredoxin. An early study of Gunsalus[33)] has, however, indicated that the flavoprotein component is replaceable, though insufficiently, by other flavoproteins. On the other hand, the second component, putidaredoxin, was found to be essential for the D-camphor hydroxylation catalyzed by P-450$_{CAM}$. Its specificity in the system was demonstrated by the general ineffectiveness of other biological electron donors in the overall hydroxylation reaction. Further details will be discussed below. The iron-sulfur protein has a molecular weight of 13,500 with two labile sulfur and two iron atoms in the molecule and therefore belongs to a class designated as 2-Fe-S* proteins.* Like many other iron-sulfur proteins of this class, the oxidized form of putidaredoxin exhibits a strong absorption band in the 280 nm region and three weaker bands in the range between 300 and 500 nm. No EPR signal attributable to the ferric iron in the oxidized putidaredoxin could be detected. This diamagnetism of the oxidized protein is believed to be due to spin coupling between two high spin ferric irons in the molecule. Upon reduction either enzymatically by the action of reductase or chemically by a reducing agent, EPR signals with g values at $g_\perp = 1.94$ and $g_{11} = 2.01$ which are typical for the reduced form of iron sulfur proteins of this class, become observable. In the optical spectra, the loss of three absorption maxima at 328, 412 and 455 nm and appearance of a new band at 550 nm were demonstrated. Available evidence obtained by EPR signal integrations, titration followed by changes in EPR, and optical spectra and magnetic susceptometry indicated that putidaredoxin undergoes a one electron transfer as do other 2-Fe-S* proteins. E_0' was shown to be -240 mV. For further discussion of the nature and molecular properties of the iron sulfur protein, readers are referred to other review articles (e.g. ref. 30).

It has been shown that the camphor complex of ferric P-450$_{CAM}$, i.e. the high spin ferric form, can be quantitatively reduced by NADH under anaerobic conditions, although the rate of reduction was extremely slow.[38)] The addition of either putidaredoxin or its reductase alone caused only a slight enhancement of the reduction rate. In contrast, the high spin ferric form in the presence of a complete electron transfer system (i.e. with both putidaredoxin and its reductase) was reduced

---

* S* is used to denote labile sulfur in accordance with Orme-Johnson et al.

instantaneously by NADH at a rate of more than 10^2–10^3 times that with the incomplete system.[30,38] In the full system without camphor, NADH was able to reduce both the flavoprotein and putidaredoxin but not the P-450$_{CAM}$. The reduction of putidaredoxin by NADH was dependent on the presence of the reductase. Thus, the electron transfer sequence shown in Fig. 3 is indicated. However, the role of putidaredoxin in the overall hydroxylation reaction appears to represent something more than just electron transfer as will be discussed below.

Putidaredoxin was found to be essential also for the formation of the reaction product, 5-*exo*-hydroxycamphor.[30,38] When the reduced form of putidaredoxin was added to either ferric, ferrous or oxy-P-450$_{CAM}$ in the presence of D-camphor, the formation of hydroxycamphor was always detectable. NADH and putidaredoxin reductase were no longer required, indicating that they were not involved directly in the hydroxylation reaction but merely served as electron donors for the P-450$_{CAM}$. On the other hand, addition of other biological electron carriers such as reduced spinach ferredoxin and adrenodoxin to the system were found to be incapable of producing the reaction product, in spite of the fact that they could reduce ferric P-450$_{CAM}$ to the ferrous state.[38] In contrast, cytochrome b_5 and rubredoxins from various sources, the redox potentials of which are known to be far above that of putidaredoxin, were able to replace the putidaredoxin, though partially. Hence, the specificity of the electron carriers for product formation could not be attributed to their redox potential. NADH and dithionite, which can reduce ferric P-450$_{CAM}$, were also inert in this respect.

The special effect of putidaredoxin described above became most apparent when the reaction between oxy-cytochrome P-450 and reduced putidaredoxin was examined.[30,36,38] As already mentioned, oxy-cytochrome P-450 with camphor present was rather stable, but underwent slow spontaneous decomposition into the high spin ferric form. The rate of decomposition was too small for it to be part of the overall hydroxylation reaction, and no detectable product formation was associated with the autodecomposition phenomenon. The addition of NADH, dithionite or iron-sulfur proteins such as ferredoxin to the oxycytochrome P-450$_{CAM}$ failed to produce hydroxycamphor in detectable quantities. On the other hand, the addition of reduced putidaredoxin to the oxy-cytochrome P-450 resulted in an instantaneous decomposition of the oxy-form, giving reaction products in accordance with the following equation.

$$\text{Ferrous P-450}_{CAM}\text{-O}_2\text{-cam} + \text{redoxin (reduced)} \longrightarrow$$
$$\text{ferric P-450}_{CAM} + \text{redoxin (oxidized)} + \text{cam-OH} + \text{H}_2\text{O} \qquad (5.2)$$

The pseudo-first order rate constant for the degradation of the oxy-form was estimated to be greater than 100 sec^{-1} in the presence of an excess of reduced putidaredoxin indicating that the degradation rate is sufficiently large for it to be part of the overall reaction.[35-37]

The oxidized form of putidaredoxin was also found to stimulate the degradation of oxy-P-450$_{CAM}$ although the rate was at least 100 times slower than that with reduced putidaredoxin.[36] The decomposition was found to be accompanied by product formation with the stoichiometry shown below.[30,38]

$$2 \text{ Ferrous P-450}_{CAM}\text{-O}_2\text{-cam} \xrightarrow{\text{oxidized redoxin}} 2 \text{ ferric P-450}_{CAM} + \text{cam-OH} + O_2 + H_2O \qquad (5.3)$$

According to Gunsalus and his co-workers, the reaction followed second order kinetics and could be explained as follows. First, oxidized putidaredoxin combines with oxy-P-450$_{CAM}$ to form a binary enzyme complex. The complex in which the reactivity of the oxy-heme is altered then receives one electron from the second oxy-P-450 molecule to degrade into the products, hydroxycamphor, H_2O and ferric P-450$_{CAM}$. The second oxy-P-450$_{CAM}$ which was in equilibrium between the oxy-form and ferrous form plus free oxygen, serves as the electron donor regenerating ferric P-450$_{CAM}$ and unbound oxygen molecule. The oxidized form of putidaredoxin is regarded as a genuine effector in this hypothesis. Kinetic experiments carried out by Tyson et al., as well as the demonstration of a stoichiometric binding of oxidized putidaredoxin with P-450$_{CAM}$, seem to support the above hypothesis.[30,38] Such an interpretation, which is analogous to the concept that the reactivity of cytochrome oxidase toward oxygen is altered by the binding of cytochrome c to the oxidase,[49,50] has recently obtained further substantiation from findings on the role of cytochrome b_5 in reconstituted liver systems containing cytochrome P-450.[51,52] A possibility still remains, however, that added putidaredoxin acts as a mere electron carrier between two molecules of oxy-P-450$_{CAM}$. Gunsalus and his co-workers[30] have found that certain dithiols such as lipoic acid and dithiothreitol could produce hydroxycamphor stoichiometrically when they were added to oxy-P-450$_{CAM}$. The monothiols so far tested have been ineffective regardless of their redox potential. Further studies on the role of putidaredoxin as an effector thus seem to be necessary.

5.5 Reaction Mechanism of P-450$_{CAM}$ and Reactive Species of Oxygen

The following reaction sequence for the monooxygenation of D-camphor catalyzed by P-450$_{CAM}$ has been proposed in conformity with the data hitherto described (Fig. 5.4).[30,53] The initial step of the overall catalysis consists of a rapid binding of camphor to free ferric P-450$_{CAM}$ with a transition of the latter from a low to a high spin state. The transition, resulting in a large elevation of the redox potential of the heme iron as well as conformational changes of the protein, renders ferric P-450$_{CAM}$ susceptible to one electron reduction to a high spin ferrous P-450$_{CAM}$-camphor complex by the action of reduced putidaredoxin as the electron donor. In the next step, molecular oxygen interacts with the reduced P-450$_{CAM}$-camphor complex giving rise to a ternary complex of the reduced cytochrome, molecular oxygen and camphor, namely oxy-P-450$_{CAM}$. The adduct of molecular oxygen to the ferrous heme can be assumed to be in equilibrium with ferric heme-superoxide

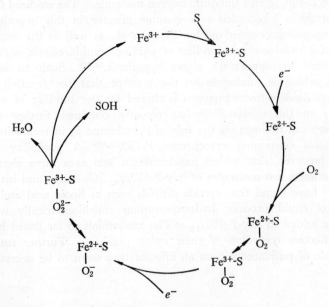

Fig. 5.4. Reaction sequence of P-450-catalyzed reactions.
(Source: ref. 53. Reproduced by kind permission of University Park Press, U.S.A.)

complex ($Fe^{3+}O_2^-$). Subsequently, the ternary complex, i.e. oxy-P-450_{CAM}, receives a second electron from the reduced putidaredoxin and degrades into reaction products through an unidentifiable reaction intermediate. A unique and important role for putidaredoxin both as an electron carrier and as an effector has been proposed to be associated with the control of product formation.

The reaction sequence described above is in accordance with the general schema proposed by Hayaishi and his co-workers for the mechanisms of oxygenases: enzyme first binds with organic substrate and then reacts with molecular oxygen to form a ternary complex of enzyme, organic substrate and oxygen (ESO_2) in which both substrates are activated. Such a mechanism was first postulated in 1964 on the basis of binding experiments[54] but the decisive evidence became available later when an ESO_2 complex of L-tryptophan 2,3-dioxygenase was demonstrated.[43,55,56] This served as the first experimental demonstration of an active enzyme-oxygen complex among many oxygen utilizing enzymes. Since then, similar ESO_2 complexes have been obtained with various kinds of oxygenases including p-hydroxybenzoate monooxygenase (a flavoprotein)[57] and cytochrome P-450 as described in this chapter. Formation of an ESO_2 complex therefore appears to be a common process to all of the oxygenases regardless of the distinction between di- and monooxygenases and differences in the nature of the prosthetic group, viz. heme, non-heme iron and flavin.

The rate determining step of cytochrome P-450-catalyzed reactions in liver microsomes has for some time been assumed to be in the transfer of the first electron to the ferric form of cytochrome P-450.[58,59] However, with a purified P-450_{CAM} system reconstituted from each of the purified components, the maximum rate of reduction has been shown to double the turnover number of P-450_{CAM} at least, as described above. Furthermore, Tyson et al.[38] were able to show that oxy-P-450_{CAM} was accumulated during the steady state of camphor hydroxylation catalyzed by a reconstituted system with the reductase, putidaredoxin, and P-450_{CAM} in the ratio 1:1:1. The ratio of these catalytic components in intact cells has been shown to be 1:1:1 by Peterson.[60] Decomposition of the oxy-P-450_{CAM} thus appears to be the rate determining process of the overall reaction. In the case of the other oxygenases cited above, the ESO_2 complexes have been shown to be so-called obligatory intermediates of the reaction and their decomposition was shown to be the rate determining process of the overall catalytic turnover.

The nature of the so-called active species of oxygen responsible for the monooxygenase reactions has been one of the most challenging problems in this field. As is generally known, molecular oxygen itself is an inert compound and does not react readily with organic substrates under ordinary conditions in the absence of enzymes. This is rather surprising since most of the oxygenase-catalyzed reactions involve the cleavage of oxygen-oxygen bonds and are therefore considerably exothermic. The reason for this low kinetic reactivity of molecular oxygen has been explained with reference to the fact that molecular oxygen has a triplet ground state whereas organic substrates are mostly in a singlet state. The direct reaction of a triplet molecule with a singlet one is a spin-forbidden process so that substantial activation energy would be required. For this reason, the reactive species of oxygen in biological systems has often been suggested to be singlet oxygen,[61,62] although this mechanism may not be operative in cytochrome P-450-catalyzed reactions (see below). On the other hand, singlet oxygen may account for the incorporation of two atoms of oxygen into organic substrates as in the case of dioxygenases. Several other candidates for the reactive oxygen in biological systems have also been proposed on the basis of chemical model experiments. A partial list includes EO_2, EOO^{2-}, EO,* $\cdot OH$, $\cdot OOH$, OH^+, and variations thereof.[63-68] However, definitive evidence for the participation of any of these intermediates in cytochrome P-450-catalyzed reactions is not available.

One important result has been described regarding the problem of oxygen activation. Guroff et al.[69] reported that during enzymatic hydroxylation of aromatic substrates, the substituent (isotopic hydrogens, halogens, methyl group) displaced by the entering hydroxyl group migrates to an adjacent position in the aromatic ring. An example of such a reaction catalyzed by liver microsomes is shown in Fig. 5.5. This intramolecular migration of the substituent associated with a hydroxylation reaction has been termed "NIH shift". Subsequently, a mechanism was proposed for the NIH shift which consisted of initial formation of arene oxide concomitant with reduction of the other half of the oxygen molecule to water.[70] Rearrangements of the arene oxide intermediate lead to the formation of a hydroxylated reaction product,

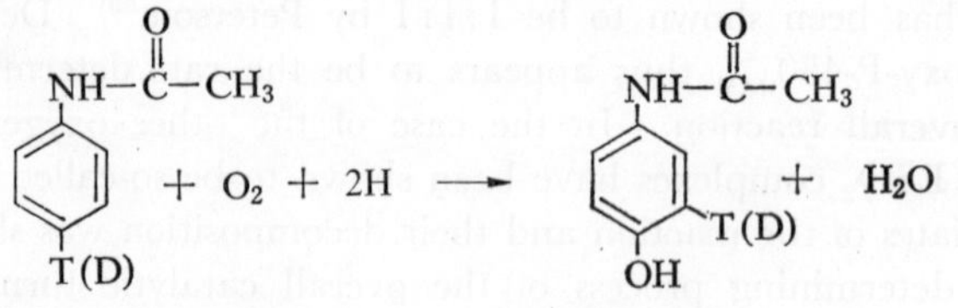

Fig. 5.5. Example of the NIH shift catalyzed by liver microsomes.

* E denotes enzyme.

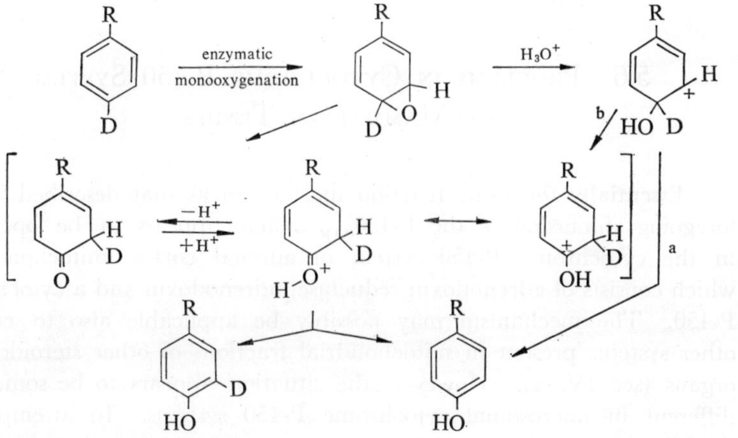

Fig. 5.6. Postulated mechanism for the hydroxylation of an aromatic compound associated with the NIH shift.[70]

although the magnitude of the substituent retention varies according to the particular pathway of rearrangement (Fig. 5.6).[70]

In support of this hypothesis, Jerina et al.[71] were able to isolate an enzymatically produced arene oxide in the naphthalene hydroxylation catalyzed by a microsomal monooxygenase system and showed it to be a reaction intermediate. This provided a significant clue to the possible nature of the reactive species of oxygen. The most likely mechanistic model for the arene oxide formation was an oxygen atom transfer process or "oxenoid mechanism" in which oxene is involved as the reactive species.[70,71] In brief, oxene is an oxygen atom with six electrons around it and is isoelectronic with carbene and nitrene. The latter two species are known to undergo various insertion and addition reactions which are analogous to the monooxygenase type of oxygen reactions. Thus, the active species of oxygen involved in certain monooxygenase reactions was postulated to be oxenoid,[70,71] revitalizing the earlier suggestions of Mason,[63] Ullrich and Staudinger,[72] and Hamilton.[73] The oxenoid reagent in the reaction cycle of P-450$_{CAM}$ described here has been suggested to be a ferric peroxide complex $(Fe^{3+}O_2^{2-})$,[70,73] in conformity with the recent finding that hydrogen peroxide and various other hydroperoxides can substitute both oxygen and two reducing equivalents for the hydroxylation of various xenobiotics catalyzed by liver microsomal systems containing cytochrome P-450.[52,74] For further details of the oxenoid theory, readers are referred to the other review articles.[70,73]

5.6 Problems in Cytochrome P-450 Systems of Mammalian Tissues

Essentially the same reaction mechanism as that described in the foregoing discussion of the P-450$_{CAM}$ system appears to be operative in the cytochrome P-450 system of adrenal cortex mitochondria[47] which consists of adrenodoxin reductase, adrenodoxin and a cytochrome P-450. The mechanism may possibly be applicable also to certain other systems present in mitochondrial fractions of other steroidogenic organs (see IV. 2). However, the situation appears to be somewhat different in microsomal cytochrome P-450 systems. In attempts to solubilize and purify an enzymatically functional cytochrome P-450 system from hepatic microsomes, Lu and Coon et al.[75–78] were able to resolve the system into three fractions, viz. cytochrome P-450, NADPH-cytochrome c reductase and a heat stable lipophilic component The reconstituted enzyme system obtained from these three components was shown to be capable of catalyzing a number of monooxygenase reactions such as those of hydrocarbon, fatty acids and of various drugs.[77,78] Phospholipids, in particular phosphatidyl choline,[79] and various detergents such as cholate, Emulgen and Triton,[52,80] were found to replace the heat stable lipophilic component. The precise role of the lipophilic component is still unknown but it appears to facilitate the interaction of the reductase with cytochrome P-450.[81,82] No other electron transferring protein, e.g. analogous to putidaredoxin, adrenodoxin or rubredoxin, has been detected in hepatic microsomes.[23,83,84]

Recent reports by Iyanagi and Mason are of special importance in this connection.[85,86] A NADPH-cytochrome c reductase purified from rabbit liver microsomes was shown to contain one mole each of FAD and FMN per single peptide of molecular weight 69,000.[86] On the basis of an extensive analysis of the redox properties of these prosthetic groups, they indicated that one flavin accepts two reducing equivalents from NADPH and the other acts as a one electron carrier between the first flavin and cytochrome P-450.[86] More recently, NADPH-cytochrome c reductase from splenic as well as kidney cortex microsomes has also been found to contain one mole each of FAD and FMN.[87]

It is well known that, besides the NADPH-linked reductase system mentioned above, hepatic microsomes contain another electron transfer system consisting of NADH-cytochrome b_5 reductase (a flavoprotein) and cytochrome b_5. The latter system has been shown by Oshino and Sato[88,89] to be functional in fatty acid desaturation. However, Estabrook et al.[90,91] have found that the addition of NADH, which preferentially reduces cytochrome b_5, enhanced the rate of monooxygenation of drugs in the presence of limited amounts of NADPH, and that a partial reoxidation of reduced cytochrome b_5 in the presence of excess NADH and organic substrate was provoked when the monooxygenase reaction was initiated by the addition of NADPH. These phenomena, termed the synergistic effect of NADH, led them to postulate an alternative electron transfer pathway for cytochrome P-450-catalyzed reactions in hepatic microsomes: the electron required for the reduction of ferric cytochrome P-450-substrate complex comes from NADPH through cytochrome c reductase, whereas the second electron is transferred from NADH via cytochrome b_5. This proposal for the specific role of cytochrome b_5 in cytochrome P-450-catalyzed reactions has been supported and further extended by recent findings of Imai and Sato[52] and hence is believed to be functional at least in the metabolism of certain drugs[51,52] (see Chapter 4).

REFERENCES

1. R. Sato, see Chapter 2 of this volume.
2. T. Omura, see Chapter 4 of this volume.
3. M. Katagiri and S. Takemori, see Chapter 4 of this volume.
4. R. Sato, *Biological and Chemical Aspects of Oxygenases* (ed. K. Bloch and O. Hayaishi), p. 195, Maruzen (1966).
5. K. Bloch and O. Hayaishi (ed.), *Biological and Chemical Aspects of Oxygenases*, Maruzen (1966).
6. K. Okunuki, M. D. Kamen and I. Sekuzu (ed.), *Structure and Function of Cytochromes*, University of Tokyo Press (1968).
7. J. R. Gillette A. H. Conney, G. J. Cosmides, R. W. Estabrook, J. R. Fouts and G. J. Mannering (ed.), *Microsomes and Drug Oxidations*, Academic Press (1969).
8. T. King, M. Morrison and H. Mason (ed.), *Oxidase and Related Redox Systems*, University Park Press (1971).
9. G. S. Boyd and R. M. S. Smellie (ed.), *Biological Hydroxylation Mechanisms*, Academic Press (1972).
10. O. Hayaishi (ed.), *Molecular Mechanisms of Oxygen Activation*, Academic Press (1974).
11. H. A. O. Hill, A. Röder and R. J. P. Williams, *Struct. Bonding*, **8**, 123 (1970).
12. T. Omura and R. Sato, *J. Biol. Chem.*, **239**, 2370, 2379 (1964).
13. M. Katagiri, B. N. Gunguli and I. C. Gunsalus, *J. Biol. Chem.*, **243**, 3543 (1968).
14. R. Tsai, C.-A. Yu, I. C. Gunsalus, J. Peisach, W. Blumberg, W. H. Orme-Johnson and H. Beinert, *Proc. Natl. Acad. Sci. U.S.A.*, **66**, 1157 (1970).
15. K. D. Williams, J. Litchfield, A. G. Miguel, T. A. van der Hoeven, D. A. Haugen, W. L. Dean and M. J. Coon, *Biochem. Biophys. Res. Commun.*, **60**, 15 (1974).
16. S. Narasimhulu, D. Y. Cooper and O. Rosenthal, *Life Sci.*, **4**, 2101 (1965).

17. Y. Imai, and R. Sato, *Biochem. Biophys. Res. Commun.*, **22**, 620 (1966).
18. H. Remer, J. B. Shenkman, R. W. Estabrook, H. Sasame, J. R. Gillette, S. Narasimhulu, D. Y. Copoer and O. Rosenthal, *Mol. Pharmacol.*, **2**, 187 (1966).
19. S. Orrenius and L. Ernster, in ref. 10. p. 215 (1974).
20. J. B. Schenkman, H. Remmer and R. W. Estabrook, *Mol. Pharmacol.*, **3**, 113 (1967).
21. W. Cammer, J. B. Schenkman and R. W. Estabrook, *Biochem. Biophys. Res. Commun.*, **23**, 264 (1966).
22. C. R. E. Jefcoate, J. L. Gaylor and R. L. Calabrese, *Biochemistry*, **8**, 3455 (1969).
23. J. Peisach and W. E. Blumberg, *Proc. Natl. Acad. Sci. U.S.A.*, **67**, 172 (1970).
24. R. W. Estabrook, J. B. Schenkman, W. Cammer, H. Remmer, D. Y. Cooper, S. Narasimhulu and O. Rosenthal, in ref. 5. p. 153 (1966).
25. M. J. Coon and A. Y. H. Lu, in ref. 7. p. 151 (1969).
26. V. Ullrich, B. Cohn, D. Y. Copper and R. W. Estabrook, in ref. 5, p. 649 (1968).
23. J. Peisach ₅yd W. E. Blumberg, *Proc. Natl. Acad. Sci. U.S.A.*, **67**, 172 (1970).
24. R. W. Estabrook, J. B. Schenkman, W. Cammer, H. Remmer, D. Y. Cooper. S. Narasimhulu, and O. Rosenthal, in ref. 5, p. 153 (1966).
25. M. J. Coon and A. Y. H. Lu, in ref. 7, p. 151 (1969).
26. V. Ullrich, B. Cohn, D. Y. Copper and R. W. Estabrook, in ref. 6, p. 649 (1968).
27. J. A. Peterson, *Arch. Biochem. Biophys.*, **144**, 678 (1971).
28. M. Sharrock, E. Münck, P. G. Debrunner, V. Marshall, J. D. Lipscomb and I. C. Gunsalus, *Biochemistry*, **12**, 258 (1973).
29. B. W. Griffin and J. A. Peterson, *ibid.*, **11**, 4740 (1972).
30. I. C. Gunsalus, J. R. Meeks, J. D. Lipscomb, D. Debrunner and E. Munck, in ref. 10, p. 559 (1974).
31. P. M. Dollinger, M. Kielczewski, J. R. Trudell, G. Barth, R. E. Linder, E. Bunnenberg, and C. Djerassi, *Proc. Natl. Acad. Sci. U.S.A.*, **71**, 399 (1974).
32. C.-A. Yu, I. C. Gunsalus, M. Katagiri, H. Suhara, and S. Takemori, *J. Biol. Chem.* **249**, 94 (1974).
33. I. C. Gunsalus, *Hoppe-Seyler's Z. Physiol. Chem.*, **349**, 1610 (1968).
34. I. C. Gunsalus, C. A. Tyson and J. D. Lipscomb, in ref. 8, p. 583 (1971).
35. Y. Ishimura, V. Ullrich and J. A. Peterson, *Biochem. Biophys. Res. Commun.*, **42**, 140 (1971)
36. Y. Ishimura, V. Ullrich and J. A. Peterson, *Fed. Proc. Fed. Am. Soc. Exptl. Biol.* **30**, 1092 (1971).
37. J. A. Peterson, Y. Ishimura, and B. W. Griffin, *Arch. Biochem. Biophys.*, **149**, 197 (1972).
38. C. A. Tyson, J. D. Lipscomb and I. C. Gunsalus, *J. Biol. Chem.*, **247**, 5777 (1972).
39. R. M. Keller, K. Wüthrich and P. G. Debrunner *Proc. Natl. Acad. Sci. U.S.A.* **69**, 2073 (1972).
40. A. E. Sidwell, Jr., R. H. Munch, E. S. G. Barron and T. R. Hogness, *J. Biol. Chem.*, **123**, 335 (1938).
41. I. Yamazaki, K. Yokota and K. Shikama, *ibid.*, **239**, 4151 (1961).
42. K. Yokota and I. Yamazaki, *Biochim. Biophys. Acta* **105**, 301 (1965).
43. Y. Ishimura, M. Nozaki, O. Hayaishi, T. Nakamura, M. Tamura, and I. Yamazaki, *J. Biol. Chem.*, **245**, 3593 (1970).
44. R. W. Estabrook, A. G. Hildebrant, J. Barbon, K. J. Netter and K. Leibman, *Biochem. Biophys. Res. Commun.*, **42**, 132 (1971).
45. P. Rösen and A. Stier, *ibid.*, **51**, 603 (1973).
46. F. P. Guengerich, D. P. Ballou and M. J. Coon, *ibid.*, **70**, 951 (1976).
47. H. Schleyer, D. Y. Cooper and O. Rosenthal, in ref. 8, p. 551 (1971).
48. T. Kimura, in ref. 5, p. 179 (1966).
49. K. Okumuki, *Oxygenases* (ed. O. Hayaishi), p. 409, Academic Press (1962).
50. Y. Orii and T. E. King, *FEBS Lett.* **21**, 199 (1972).
51. A. Y. H. Lu, W. Levin, H. Selander and D. M. Jerina, *Biochem. Biophys. Res Commun.*, **61**, 1348 (1974).

52. Y. Imai and R. Sato *ibid.*, **75**, 420 (1977).
53. J. A. Peterson, Y. Ishimura, J. Baron and R. W. Estabrook, in ref. 8, p. 565 (1971).
54. O. Hayaishi, Proc. Plenary Sessions, 6th Int. Congr, Biochemistry, vol. 33, p. 31 (1964).
55. Y. Ishimura, M. Nozaki, O. Hayaishi, M. Tamura and I. Yamazaki, *J. Biol. Chem.*, **242**, 2574 (1967).
56. O. Hayaishi, Y. Ishimura, H. Fujisawa and M. Nozaki, in ref. 8, p. 125 (1971).
57. T. Spector and V. Massey, *ibid.*, **247**, 5632 (1972).
58. P. L. Gigon, T. E. Gram and J. R Gillette *Mol. Pharmacol.*, **5**, 109 (1969).
59. D. S. Davies, P. L. Gigon and J. R. Gillette, *Life Sci.*, **8**, part II, 85 (1969).
60. J. A. Peterson, *J. Bact.*, **103**, 714 (1970).
61. A. H. Soloway, *J. Theoret. Biol.* **13**, 100 (1966).
62. C. S. Foote, *Accounts Chem. Res.*, **1**, 136 (1968).
63. H. S. Mason, *Advan. Enzymol.*, **19**, 79 (1957).
64. S. Udenfriend, C. T. Clark, J. Axelrod and B. B. Brodie, *J. Biol. Chem.*, **208**, 731 (1954).
65. R. R. Grinstead, *J. Am. Chem. Soc.*, **82**, 3472 (1960).
66. Hj. Staudinger and V. Ullrich, *Z. Naturforsch.*, **196**, 409, 877 (1964).
67. R. O. C. Norman and J. R. Lindsay-Smith, (ed. T. E. King, H. S. Mason and M. Morrison), *Oxidases and Related Redox Systems*, vol. I, p. 131, John Wiley (1965).
68. C. Chen and C. C. Lin, *Biochim. Biophys. Acta.* **170**, 366 (1968).
69. G. Guroff, J. W. Daly, D. M. Jerina, J. Renson, B. Witkop and S. Udenfriend, *Science*, **157** 1524 (1967).
70. D. M. Jerina and J. W. Daly, in ref. 8, p. 143 (1971).
71. D. M. Jerina, J. W. Daly, B. Witkop. P. Zaltzman-Nirenberg and S. Udenfriend, *Biochemistry*, **9**, 147 (1970).
72. V. Ullrich and Hj. Staudinger, in ref. 5, p. 235 (1964).
73. G. A. Hamilton, *J. Am. Chem. Soc.*, **86**, 3391 (1964).
74. G. D. Nordblom, R. E. White and M. J. Coon *Arch. Biochem. Biophys.*, **175**, 524 (1976).
75. A. Y. H. Lu and M. J. Coon, *J. Biol. Chem.*, **243**, 1331 (1968).
76. A. Y. H. Lu, K. W. Junk and M. J. Coon. *ibid.*, **244**, 3714 (1969).
77. A. Y. H. Lu, H. W. Strobel and M. J. Coon, *Biochem. Biophys. Res. Commun.*, **36**, 545 (1969).
78. A. Y. H. Lu, H. W. Strobel and M. J. Coon, *Mol. Pharmacol.*, **6**, 213 (1970).
79. H. W. Strobel, A. Y. H. Lu, J. Heidema and M. J. Coon, *J. Biol. Chem.*, **245**, 4851 (1970).
80. A. Y. H. Lu, W. Levin and R. Kuntzman, *Biochem. Biophys. Res. Commun.*, **60**, 266 (1974).
81. M. J. Coon, A. P. Autor, R. F. Boyer, E. T. Lode and H. W. Strobel, in ref. 8, p. 529 (1971).
82. S. Orrenius and L. Ernster, in ref. 9, p. 215 (1974).
83. Y. Miyake, H. S. Mason and W. Landgraf, *J. Biol. Chem.*, **242**, 393 (1967).
84. Y. Ichikawa and T. Yamano, *Biochim. Biophys. Acta.*, **200**, 220 (1970).
85. T. Iyanagi and H. S. Mason, *Biochemistry*, **12**, 2297 (1973).
86. T. Iyanagi and H. S. Mason, *ibid.*, **13**, 1701 (1974).
87. T. Iyanagi, *FEBS Lett.*, **46**, 51 (1974).
88. N. Oshino, Y. Imai and R. Sato, *J. Biochem. (Tokyo)*, **69**, 155 (1971).
89. N. Oshino and R. Sato, *ibid.*, **69**, 169 (1971).
90. B. Cohen and R. W. Estabrook, *Arch. Biochem. Biophys.*, **143**, 46 (1971).
91. A. G. Hildebrandt and R. W. Estabrook, *ibid.*, **143**, 66 (1971).

Index

A
absolute extinction coefficient 102
adrenal cortex microsomes 5, 12, 80
adrenal cortex mitochondria 9, 75, 80, 173
  P-450 in— 11, 96, 124, 148
adrenodoxin 9, 27, 165
  absorption spectrum 167
  amino acid sequence 168
  iron content 168
  labile sulfide 168
  reductase — complex 172
  —-dependent reduction 171
  — reductase 27, 170
n-alkane methyl hydroxylation 189
alkyl isocyanides 89
ω-amino-n-octyl Sepharose 4B 42, 43, 46
aniline hydroxylation 68
apo-adrenodoxin 166
arene oxide 18
azo-compound 160

B
Bacillus megaterium 25, 31
bacterial oxygenase system 184
bacterial P-450 25, 190
  amino terminus 50
  carboxyl terminus 50
  conversion to P-420 62
  hydroxylase system 192
  isoelectric point 50
  molecular weight 48
bovine adrenodoxin 168

C
Caldariomyces 184
Caldariomyces fumago 8, 191
camphor 31
  hydroxylation of — 186
  — 5-*exo* hydroxylase 186
  — hydroxylase system 9
  — methylene hydroxylating system 186
  — oxygenase system 148
Candida tropicalis 25, 31, 184, 190, 195

alkane-oxidizing enzyme system 203
  electron transport system 200, 201
  microbody 203
carcinogen 32
carcinogenic compound 17
charge-transfer band 95
chemical carcinogenesis 18
chick kidney mitochondria 181
chloroperoxidase 8, 191
cholesterol 29
  7 α-hydroxylase 17
cinnamic acid 30
Claviceps purpurea 25
CO
  —-binding pigment 74, 139
  —-difference spectrum 101
  —-shift of Soret band 82
concentration difference spectrum 84
conversion of P-450 to P-420 54, 62, 74, 125, 152
corpus luteum 99, 181
Corynebacterium sp. 8, 25, 31, 184, 189
cyanide-sensitive factor 140, 195
cytochrome b_1 194
cytochrome b_5 74, 150, 154, 194, 225
  amphipatic nature 143
  content in microsomes 142
  discovery 138
  g value 108
  selective removal 12, 76
cytochrome cc' 61
cytochrome P-420→P-420
cytochrome P-450→P-450

D
deoxycorticosterone 180
desaturation reaction of palmitoyl CoA 195
detergent 6
drug-oxidizing enzyme system 159

E
Echinocystis macrocarpa endosperm 8
electric configuration 111
electron transfer 145

Emulgen 41, 43
EPR signal
 P-450 118
 P-420-NO complex 121
 P-450-NO complex 121
 phenylisocyanide-P-450 120
EPR spetrcum
 liver microsome 116
 P-450$_{CAM}$ 122
 rabbit liver 107
ergosterol 29, 207
Ergot alkaloid 30
ethyl isocyanide 62, 74
——complex 99
——difference spectrum 62, 66, 88, 89
extinction coefficient 100

F
ferric hemoprotein 81
ferric high spin-type 82
ferric low spin-type 77, 82
ferrihemochrome type 82
ferriprotoporphyrin IX 47
ferrohemochrome type 82
ferrous high spin-type 77, 82
ferrous low spin-type 82
fetal liver 33
fluid-like structure of membrane 144

G
gibberellin 30
glycerol 80

H
hemepeptide 51
hepatic microsomal oxygenase system 138
hepatic microsomal P-450 94, 155
 molecular weight 48
 purification 37
hepatomas 33
higher plant 25
high spin band 95
high spin heme 110
high spin-low spin difference spectrum 93
high spin-type absorption spectrum 76
high spin-type P-450 98, 115
horse myoglobin 129
horseradish peroxidase 129
hydrophobic environment 59
hydroxylation
 aromatic compound 223
 7α—— 158
 15β—— 158
 ω—— 30, 152, 189

I
immunochemical study 147
integral membrane protein 142
iron-sulfur protein 216

L
ligand field theory 110
liver microsomes 73, 156
 CO-difference spectrum 2, 89
 electron transport system 138
 ESR 4
 hydroxylation of testosterone 156
 MC-treated rabbit 45
 MC-treated rat 40, 41
 methyl-cholanthrene-induced 76
 NADH difference spectrum 89
 NADPH-cytochrome c reductase activity 139
 PB-treated rabbit 40-44
 PB-treated rat 40, 41
 phenobarbital-induced 76
 titration 158
low spin heme 110
low spin type absorption spectrum 76
Lubrol WX 41

M
magnetic circular dichroism (MCD) 52, 132
mercaptide anion 83
methemoglobin-thioalcohol complex, g value 119
3-methylcholanthrene 14, 24
metyrapone 123, 177
microbial P-450 184
microsomal CO-binding pigment 2, 73
—— drug-oxidizing activity 6, 14
—— electron transport system 141, 148
 EPR 107
—— Fe$_X$ 4, 51, 106, 126, 139
—— NADH-linked electron transport system 140
 oxidation-reduction potential 109
—— oxygenase system 10, 147
—— P-450 80
—— signal 79, 109
microsomes 24, 75, 138
mitochondria 24
mitochondrial P-450 79
mixed spin-type P-450 97
modified type II difference spectrum 90
molar extinction coefficient 100
mold 25
molecular aggregate 143

molecular architecture of microsomal membrane 142
Mössbauer spectrum 131

N

NADH-cytochrome b_5 reductase 150, 194, 224
  Amphipatic nature 143
  content 142
  discovery 139
NADH-dependent drug oxidation system 150
NADH-putaredoxin reductase 188
NADH-putidaredoxin reductase 186, 188, 216
NADH-rubredoxin reductase 189
NADH synergism 150
NADPH-adrenodoxin reductase 27, 170
  absorption spectrum 172
  flavin 172
  molecular weight of— 172
  property 171
  purification 170
  ultracentrifugation pattern 171
NADPH-cytochrome c reductase 27, 142, 143, 224
  discovery 139
  purification 153
  solubilization 153
NADPH-dependent drug oxidation reaction 150
NADPH-P-450 reductase 140, 194
natural coordinant 83
NIH shift 222
nitro-compound 160
nitrogen monoxide complex 99
NMR 127
nuclear envelope 24

O

n-octane 31
opaque glass method 76
orbital energy difference 113
organic hydroperoxide 28
oxenoid mechanism 229
oxygen
oxygenated hemoprotein 215
oxygen interaction with P-450 214
  reactive species 220, 222
oxy-P-450 89, 100, 218
oxy-P-450$_{CAM}$ 215, 219

P

P-420 38
  absolute spectrum 100
  CO-difference spectrum 3
  multiplicity 60

  purification 38
P-448 15, 85, 157
P-450
  absolute absorption spectrum 11, 75, 94
  amino acid composition 48
  axial ligand 50
  biphasic reduction 146
  carbohydrate component 50
  CD spectrum 61
  chemical property 47
  CO-complex 101
  CO-difference spectrum 3, 84
  content in microsomes 142
  conversion to P-420 4
  difference spectrum 83
  discovery 1, 139
  distribution 7, 23
  EPR 106, 118, 123
  ethyl isocyanide-difference spectrum 3, 62
  evolutionary aspect 18
  extinction coefficient 4, 100
  facultatively aerobic yeast cell 204
  g value 120
  high spin EPR 115
  history 1
  homogeneous preparation 42
  hydrophobic environment 61
  immunochemical property 53
  interaction with isocyanide 62
  isocyanide-difference spectrum 87
  low temperature spectrum 99
  magnetic property 106
  magnetic susceptibility 131
  MCD spectrum 132
  methylcholanthrene-induced — 157
  molecular property 10, 37
  molecular weight 13, 47
  Mössbauer spectrum 131
  multiple form 14, 26, 157
  native conformation 61
  NMR 127
  optical property 73
  orbital energy 120
  ORD 57
  oxidation-reduction cycle 66
  oxidation-reduction potential 13, 52
  oxidized form 65
  oxygenated form 215
  phenobarbital-induced — 157
  photochemical action spectrum 102
  photo-irradiation of CO-compound 70

physiological function 5
prosthetic group 47
purification 43–45
reduced form 62
reduced minus oxidized difference spectrum 89
reduction 149
reductive function 159
solubilization 38, 40, 46
spectral abnormality 81
stabilization 11
substrate-binding specificity 91
substrate-binding spectrum 13
substrate-induced difference spectrum 90
—-catalyzed reaction 209, 212, 220, 221
—-containing oxygenase system 8
— content of yeast cell 204
— heme ligand 113
— particle 39, 77
— preparation 79
P_1-450 15, 85, 157
EPR 117
$P-450_{CAM}$ 80, 184, 186, 210
absolute spectrum 95
absorption spectrum 187, 215
binding with D-camphor 213
E_0' 213
electron transfer system 216
EPR 122
function 188
molecular activity 213
NMR 127
optical property 212
purification 186
reaction mechanism 220
resonance Raman spectrum 132
substrate binding 187
—-substrate complex 188, 189
$P-450_{CB}$ 184, 191
absorption spectrum 191
$P-450_{SCC}$
absorption spectrum 176
electrophoresis 178
g value 125
molecular weight 48, 178
property 175
purification 173
$P-450_{11\beta}$ 180
absorption spectrum 176
electrophoresis 178
g value 125
molecular weight 48, 178

property 175
purification 173
$P-450_\omega$ 185
Penicillium patulum 25
peroxidation 86
phenobarbital 14, 24
phenobarbital-induced liver microsome 76
phosphatidyl choline 152
photochemical action spectrum 5, 102
photo-irradiation 70
placenta 181
polychlorinated biphenyls 26
protecting agent 39
Pseudomonas oleovorans 189
Pseudomonas putida 8, 9, 25. 31, 80, 184, 186
putidaredoxin 9, 27, 186, 188, 210, 216, 218, 219
E_O' 213
— reductase 27, 210
R
rabbit antibody against P-450 177
reconstruction
microsomal oxygenase system 147
oxygenase activity 151
reconversion
P-420 to P-450 60
reduced minus oxidized spectrum 89
resonance Raman (RR) spectroscopy 132
Rhizobium japonicum 25, 80, 184, 190
— bacteroid 7, 31
— P-450 99, 117, 190
— $P-450_C$ 100
rubredoxin 189
S
Saccharomyces cerevisiae 7, 25, 195, 204
— NADPH-cytochrome c reductase 200
— P-450
absorption spectrum 196
difference spectrum 198
purification 196
spectral parameters 197
Sepharose 4B 42, 43, 46
sequential fragmentation 165
side-chain cleavage 177, 180
singlet oxygen 222
sodium cholate 126
Solghum bicolor seedling 8
solvent difference spectrum 84
spironolactone 177

stearyl coenzyme A desaturase 140
steroid hormone 30
steroid 11β-hydoxylase system 9
steroid 21-hydroxylation 29
  photochemical action spectrum 102
sterol biogenesis 201
Streptomyces erythreus 25
substrate-binding spectrum 12
—-induced difference spectrum 80, 90
— interaction with P-450 211
superoxide anion 169
synergistic effect of NADH 225

T

terminal oxidase 103
tertiary amine *N*-oxide 29, 160
testis 181
tetradecane 31
thiamine disulfide 119
thiolate anion 50, 61, 83
Triton N-110 40, 41
type I difference spectrum 81, 90
type I spectral change 146
type II difference spectrum 81, 90
type II spectral change 146

V · W

Vinca rosea seedling 8
vitamin D_3 30
vitamin K_3-dependent NADPH oxidase 200
Warburg's respiratory enzyme 103

X · Y

xenobiotics 31
yeast 25
— electron transport system 200, 201
— microsomal electron transport system 194
— microsomal oxygenase system 194
— P-450 99, 195
  conversion to P-420 199
  disappearance 202, 206
  induction 202, 206
  molecular weight 198
  property 196
  spin state 196
  spectral change 199

WITHDRAWN
FROM STOCK
QMUL LIBRARY